Differential Equations
A Maple™ Supplement
Second Edition

Textbooks in Mathematics

Series editors:
Al Boggess, Kenneth H. Rosen

Real Analysis
With Proof Strategies
Daniel W. Cunningham

Train Your Brain
Challenging Yet Elementary Mathematics
Bogumil Kaminski, Pawel Pralat

Contemporary Abstract Algebra, Tenth Edition
Joseph A. Gallian

Geometry and Its Applications
Walter J. Meyer

Linear Algebra
What You Need to Know
Hugo J. Woerdeman

Introduction to Real Analysis, Third Edition
Manfred Stoll

Discovering Dynamical Systems Through Experiment and Inquiry
Thomas LoFaro, Jeff Ford

Functional Linear Algebra
Hannah Robbins

Introduction to Financial Mathematics
With Computer Applications
Donald R. Chambers, Qin Lu

Linear Algebra
An Inquiry-based Approach
Jeff Suzuki

Mathematical Modeling in the Age of a Pandemic
William P. Fox

Differential Equations: A Maple™ Supplement, Second Edition
Robert P. Gilbert, George C. Hsiao, Robert J. Ronkese

https://www.routledge.com/Textbooks-in-Mathematics/book-series/
CANDHTEXBOOMTH

Differential Equations
A Maple™ Supplement
Second Edition

Robert P. Gilbert
George C. Hsiao
Robert J. Ronkese

CRC Press
Taylor & Francis Group
Boca Raton London New York

CRC Press is an imprint of the
Taylor & Francis Group, an **informa** business

A CHAPMAN & HALL BOOK

Second edition published 2021
by CRC Press
6000 Broken Sound Parkway NW, Suite 300, Boca Raton, FL 33487-2742

and by CRC Press
2 Park Square, Milton Park, Abingdon, Oxon, OX14 4RN

© 2021 Robert P. Gilbert, George C. Hsiao, Robert J. Ronkese

First edition published by Pearson Education Inc, 2002

CRC Press is an imprint of Taylor & Francis Group, LLC

Library of Congress Cataloging-in-Publication Data

Names: Gilbert, Robert P., 1932- author. | Hsiao, G. C. (George C.),
author. | Ronkese, Robert J., author.
Title: Differential Equations: A Maple™ Supplement, Second Edition / Robert P.
Gilbert, George C. Hsiao, Robert J. Ronkese.
Description: Second edition. | Boca Raton : Chapman & Hall/CRC Press, 2021.
| Series: Textbooks in mathematics | Includes bibliographical references
and index.
Identifiers: LCCN 2021000955 (print) | LCCN 2021000956 (ebook) | ISBN
9781032007816 (paperback) | ISBN 9781032021799 (hardback) | ISBN
9781003175643 (ebook)
Subjects: LCSH: Differential equations--Data processing. | Maple (Computer
file)
Classification: LCC QA371.5.D37 G54 2021 (print) | LCC QA371.5.D37
(ebook) | DDC 515/.350785536--dc23
LC record available at https://lccn.loc.gov/2021000955
LC ebook record available at https://lccn.loc.gov/2021000956

ISBN: 978-1-032-02179-9 (hbk)
ISBN: 978-1-032-00781-6 (pbk)
ISBN: 978-1-003-17564-3 (ebk)

DOI: 10.1201/9781003175643

Typeset in CMR10
by KnowledgeWorks Global Ltd.

Contents

1 **Introduction to the Maple DEtools**   **1**
   1.1   Analytical Solutions and Their Plotting 1
   1.2   Direction Fields and Integral Curves 3
   1.3   Computer Lab 8
   1.4   Supplementary Maple Programs 9
      1.4.1   Implicit and Explicit Solutions 9
      1.4.2   Numerical Solutions 9

2 **First-Order Differential Equations**   **11**
   2.1   Linear Differential Equations 11
   2.2   Project: Mixing Problems 18
      2.2.1   Project1: One Tank 18
      2.2.2   Project2: Two Tanks 23
   2.3   Separable Differential Equations 24
   2.4   Exact Equations 27

3 **Numerical Methods for First-Order Equations**   **33**
   3.1   Picard's Iteration Method and Semi-Batch Reactor 33
   3.2   An Existence and Uniqueness Theorem 33
   3.3   Picard Iteration Method 34
   3.4   Computer Lab 39
   3.5   Numerical Procedures and Fermentation Kinetics 41
   3.6   The Euler Method 42
   3.7   Higher-Order Methods 43
   3.8   Maple Procedures 43
   3.9   Computer Lab 50
   3.10 Supplementary Maple Programs 51
      3.10.1 The Order of Convergence 51

4 **Differential Equations with Constant Coefficients**   **55**
   4.1   Second-Order Equations with Constant Coefficients 55
   4.2   Variation of Parameters 57
      4.2.1   The Wronskian 57
   4.3   The Method of Undetermined Coefficients 63
   4.4   Higher-Order, Homogeneous Equations 70
      4.4.1   Polynomial Solutions 71

  4.5 Nonhomogeneous Linear Equations   72
       4.5.1 Undetermined Coefficients   72
       4.5.2 Variation of Parameters   75
       4.5.3 Further Remarks on the Variation of Parameters
              Method   79

5 Applications of Second-Order Linear Equations   83
  5.1 Simple Harmonic Motion   83
  5.2 General Solutions   83
  5.3 Method of Undetermined Coefficients   85
  5.4 Additional Useful Commands   86
  5.5 Computer Lab   88
  5.6 Supplementary Maple Programs   88
       5.6.1 The Phenomenon of Beats   88
       5.6.2 The Phenomenon of Resonance   90
  5.7 Particular Solutions   91
  5.8 Computer Lab   92
  5.9 Supplementary Maple Programs   93
       5.9.1 Resonance Curves   93
       5.9.2 An Example   96

**6 Two-Point Boundary Value Problems, Catalytic Reactors
    and Boundary-Layer Phenomena   99**
  6.1 Analytical Solutions   99
  6.2 Finite-Difference Methods 101
       6.2.1 Finite-Difference Procedure for the Two-Point BVP: . 103
  6.3 Computer Lab 105
  6.4 Supplementary Maple Programs 107
       6.4.1 An Exact Asymptotic Expansion 107

7 Eigenvalue Problems   111
  7.1 Sturm-Liouville Problems 111
  7.2 Numerical Approximations 112
  7.3 The Newton-Raphson Method 118
  7.4 Computer Lab 120
  7.5 Supplementary Mapple Programs 120
       7.5.1 An Eigenvalue Equation 120

8 Power Series Methods for Solving Differential Equations 125
  8.1 Nonlinear Differential Equations 129
  8.2 Regular-Singular Points 132
  8.3 Programs for Finding Solutions 134
  8.4 Projects 146

9 Nonlinear Autonomous Systems **147**
9.1 The Taylor Series Method 147
9.2 The Phase Plane 148
9.3 Linear Systems 149
9.4 Useful Maple Commands 157
9.5 Computer Lab 157
9.6 Supplementary Maple Programs 158
   9.6.1 Taylor Series Expansion 158
   9.6.2 The Damped Pendulum 162

10 Integral Transforms **167**
10.1 The Laplace Transform of Elementary Functions 167
10.2 Solving Differential Equations with the Laplace Transform . 172
10.3 Fourier Transforms 179

11 Partial Differential Equations **191**
11.1 Elementary Methods 191
11.2 The First-Order Partial Differential Equation 194
11.3 The Heat Equation 199
11.4 The Vibrating String 205
   11.4.1 Separation of Variables with MAPLE. 208
11.5 The Laplace Equation 211

12 Transmutations **219**
12.1 The Method of Ascent 219
12.2 Orthogonal Systems of Functions 226
12.3 Acoustic Propagation 227

Bibliography **231**

Index **233**

Chapter 1

Introduction to the Maple DEtools

The purpose of this chapter is to introduce some Maple features specially related to differential equations. In some instances Maple can find an analytical expression for the solution of an ordinary differential equation (ODE). For this purpose we need to introduce certain **syntax** which will be given in this chapter.

1.1 Analytical Solutions and Their Plotting

Let us consider the first-order differential equation,

$$(E) \qquad \frac{d\,y(x)}{dx} = f(x,y).$$

To obtain the *general solution* for a particular form of the above equation, we first load **DEtools** as follows:

> `with(DEtools):`

Let us chose the right-hand side of the above equation [1] as

> `f:=2*x*y(x)+x;`

$$f := 2\,x\,y(x) + x$$

The **syntax** for entering and solving the differential equation with this right-hand-side are then given as

> `deq:=diff(y(x),x)=f;`

$$deq := \frac{\partial}{\partial x}\,y(x) = 2\,x\,y(x) + x$$

> `gsoln:=dsolve(deq,y(x));`

$$gsoln := y(x) = -\frac{1}{2} + e^{(x^2)}\, \_C1$$

---

[1] Notice that Maple requires the unknown function to be inputted as $y(x)$.

To plot the solutions all parameters must be given as **floating point** numerical values. First we must **load plots**. Notice that the use of the colon at the end of the input line suppresses the output from this line.

> `with(plots):`

For ODEs with *side conditions*, such as an **initial value condition**, abreviated as (*IVC*), one first **defines** the condition as

> `ic:=y(0)=y0;`

$$ic := \mathrm{y}(0) = y0$$

> `soln:=dsolve({deq,ic},y(x));`

$$soln := \mathrm{y}(x) = -\frac{1}{2} + e^{(x^2)}\left(\frac{1}{2} + y0\right)$$

We then define the function to be plotted as the right-hand-side of `soln`, and then substitute in the initial condition $y(0) = 3$.

> `phi:=rhs(soln);`
> `phi1:=subs({y0=3},phi);`

$$\varphi 1 := -\frac{1}{2} + \frac{7}{2}e^{(x^2)}$$

To plot the graph of $\varphi_1(x)$ from $x = 0$ to $x = 10$ we use the syntax

> `plot(phi1,x=0.1);`

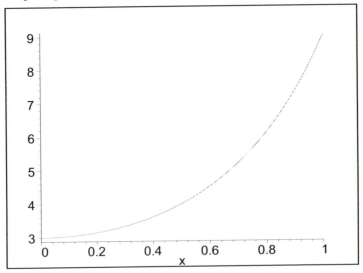

1.2   Direction Fields and Integral Curves

The slope $\frac{dy(x)}{dx}$ of equation(E) is given by $f(x, y(x))$. If at each point (x, y) we draw a short line segment in the direction of the slope $f(x, y)$ we generate the **direction field** of the differential equation. Using Maple we may plot the *direction field* for a differential equation without actually solving it. We do this by using the command `dfieldplot`:

```
>   with(DEtools):
>   deq:=diff(y(x),x)=2*x*y(x)+x;
```

$$deq := \frac{\partial}{\partial x} y(x) = 2\, x\, y(x) + x$$

```
>   dfieldplot(deq,y(x),x=-1.1,
>   y=-2.10);
```

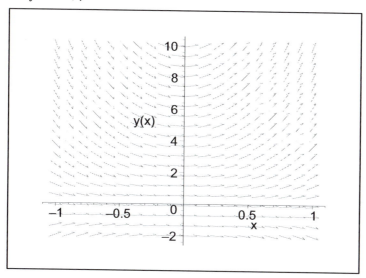

For **integral curves**, i.e. the particular solutions to an **initial value problem** we use the command `DEplot` to plot these solutions, without actually solving the differential equations. First let us introduce a set of initial conditions as

```
>   inits:=[y(1)=-1],[y(1)=0],[y(1)=2];
```

$$inits := [y(1) = -1],\ [y(1) = 0],\ [y(1) = 2]$$

We then plot the corresponding solutions as

```
>   DEplot(deq,y(x),x=0.1,inits);
```

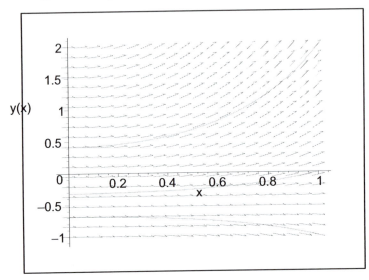

To superimpose the *direction fields* and the *exact solution* , we use the command `display`:

```
>   p1:=plot(phi1,x=-2.2):
>   p2:=dfieldplot(deq,y(x),
>   x=-2.2,y=-2.10):
```

We have introduced `display` because it gives us the possibility to simultaneously plot several functions, not necessarily solutions of differential equations, on the same screen.

```
>   display({p1,p2});
```

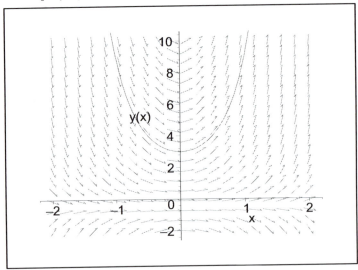

```
> DEplot(deq,y(x),x=-2.2,inits,y=-2.2);
```

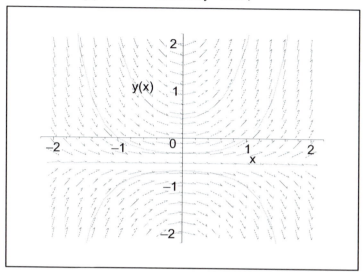

We now present as a Maple session the treatment of the differential equation

$$\frac{d\,y(x)}{dx} = x - y(x)$$

and run through the details, showing both input and output lines, from beginning to end. First we must load

```
> with(DEtools):
```

Then we input the differential equation

```
> f:=x-y(x);
```

$$f := x - y(x)$$

```
> dfieldplot(D(y)(x)=f,y(x),x=-3.3,y=-3.3);
```

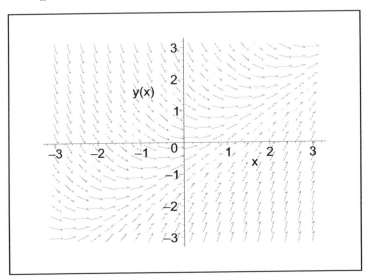

```
>  deq:=diff(y(x),x)=f;
```

$$deq := \frac{\partial}{\partial x} y(x) = x - y(x)$$

```
>  soln:=dsolve(deq,y(x));
```

$$soln := y(x) = x - 1 + e^{(-x)} \_C1$$

Maple can actually produce the analytic solution in this case. Indeed, we obtain

```
>  phi:=rhs(%);
```

$$\varphi := x - 1 + e^{(-x)} \_C1$$

where _C1 designates an arbitrary constant.

```
>  with(plots):
>  plot({subs(_C1=4,phi),
>  subs(_C1=2,phi),(_C1=1,phi),subs(_C1=0,phi),
>  subs=(_C1=-1,phi),subs(_C1=-2,phi),subs=(_C1=-4,phi)
>  },x=-3.3,y=-3.3);
```

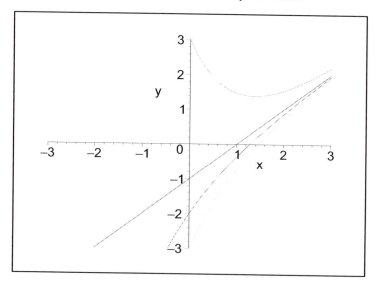

```
>  inits:=
>  {[y(0)=-6],[y(0)=-3],[y(0)=-2],[y(0)=-1],[y(0)=0],
>  [y(0)=1],[y(0)=2],[
>  y(0)=3]};
```

$$inits := \{[y(0) = -6], [y(0) = -3], [y(0) = -2], [y(0) = -1], [y(0) = 0],$$

$$[y(0) = 1], [y(0) = 2], [y(0) = 3]\}$$

```
>  p1:=DEplot(diff(y(x),x)=f,y(x),
>  x=-3.3,y=-3.3,inits):
>  display(p1);
```

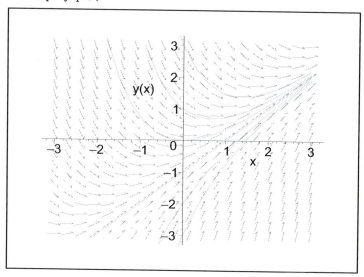

1.3   Computer Lab

The liquid level h in a tank emptying through an opening at its base by virtue of the pressure on the liquid due to gravity may be modeled by the first-order ODE:

$$A\frac{dh}{dt} = -k\sqrt{h}$$

where A is the cross-sectional area of the tank and k is a constant characterizing the opening through which the fluid flows.

- Use Maple to solve this equation, subject to the initial condition:

$$h = h_0 \quad \text{at} \quad t = 0.$$

- Specify the values of A, h_0, k as 12, 10.75, 5.2 and plot the solution in the range from $t = 0$ to $t = $ some upper limit, say t_{max}. From the graph, find the time it takes to empty the tank.

  Hint: one may use the command subs to substitute the specific data:

  ```
  >par1:={A=12,h0=10.75,k=5.2};
  >soln1:=subs(par1, phi);
  ```

  Note: phi denotes the right hand side of the solution in terms of the parameters A, h_0, k.

- Compare your result with the time to empty obtained from the Maple analytic solution via:

  ```
  >solve(soln1=0, t);
  ```

  Note: soln1 denotes your solution above with the specified values of A, h_0, k.

- Explore the effect on the solution of using values of the parameters. Explain in physical terms what happens when k is increased, A is increased, Yes and h_0 is increased.

Note that when you are done, A, k, etc. retain the numerical values you have given them. To clear these values so that the symbols revert to being simply algebraic symbols, use the commands: k:='k'; etc. We will see in a subsequent session that we may avoid the necessity to undefine parameters by simply using **local variables**.

1.4 Supplementary Maple Programs

1.4.1 Implicit and Explicit Solutions

In the following, implicit and explicit solution forms are given:

Chemical Reactions

```
>   with(DEtools):
>   deq:=diff(x(t),t)=alpha*(q-x(t))*(p-x(t));
```

$$deq := \frac{\partial}{\partial t} x(t) = \alpha \left(q - x(t)\right) \left(p - x(t)\right)$$

```
>   impsoln:=dsolve(deq,x(t),implicit);
```

$$impsoln := \frac{\ln(q - x(t))}{\alpha \left(-q + p\right)} - \frac{\ln(p - x(t))}{\alpha \left(-q + p\right)} + t + \_C1 = 0.$$

As you see, this solution is only valid when the solution $x(t)$ is smaller than both p and q. This example shows that one should not trust Maple blindly!!! Another example of this sort will be given later.

```
>   expsoln:=solve(%,x(t));
```

$$expsoln := -\frac{p - q\, e^{(-t\,\alpha\,q + t\,\alpha\,p + \_C1\,\alpha\,q - \_C1\,\alpha\,p)}}{e^{(-t\,\alpha\,q + t\,\alpha\,p + \_C1\,\alpha\,q - \_C1\,\alpha\,p)} - 1}$$

```
>   gsoln:=dsolve(deq,x(t),explicit);
```

$$gsoln := x(t) = -\frac{p - q\, e^{(-t\,\alpha\,q + t\,\alpha\,p + \_C1\,\alpha\,q - \_C1\,\alpha\,p)}}{e^{(-t\,\alpha\,q + t\,\alpha\,p + \_C1\,\alpha\,q - \_C1\,\alpha\,p)} - 1}$$

Population Dynamics

```
>   ode:=diff(p(t),t)=(b-c*p(t))*p(t);
```

$$ode := \frac{\partial}{\partial t} p(t) = (b - c\, p(t))\, p(t)$$

```
>   soln:=dsolve({ode,p(0)=P_0},p(t));
```

$$p(t) = \frac{P\_0\, b}{P\_0\, c + e^{(-b\,t)}\, b - e^{(-b\,t)}\, P\_0\, c}$$

1.4.2 Numerical Solutions

Assuming we have already performed with(Detools) define the function f.

```
>  with(DEtools):with(plots):
```

Warning, the name changecoords has been redefined

```
>  f:=(x,y)->y^2;
```

$$f := (x, y) \rightarrow y^2$$

```
>  deq2:=diff(y(x),x)=f(x,y(x));
```

$$deq2 := \frac{\partial}{\partial x} \, y(x) = y(x)^2$$

```
>  inc:=y(0)=1;
```

$$inc := y(0) = 1$$

```
>  soln2:=dsolve({deq2,inc},y(x));
```

$$soln2 := y(x) = -\frac{1}{x - 1}$$

```
>  soln3:=dsolve({deq2,inc},y(x),numeric);
```

$$soln3 := \mathbf{proc}(rkf45\_x) \ldots \mathbf{end\ proc}$$

```
>  soln3(.99);
```

$$[x = .99, \, y(x) = 100.0000081873320]$$

```
>  f1:=plot(rhs(soln2),x=0..99):
>  g2:=[seq([rhs(soln3(.1*k)[1]),rhs(soln3(.1*k)[2])],k=0..9)];
```

$$g2 := [[0., 1.], [.1, 1.11111211247547770], [.2, 1.25000109119219038],$$
$$[.3, 1.42857157233628351], [.4, 1.66666683037201046],$$
$$[.5, 2.00000243471640138], [.6, 2.49999964404530894],$$
$$[.7, 3.33333480122293624], [.8, 5.00000544742327158],$$
$$[.9, 9.9999970980296418]]$$

```
>  f2:=plot(g2,x=0..1,style=POINT,symbol=CIRCLE):
>  display({f1,f2});
```

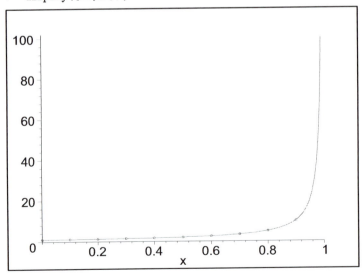

Chapter 2

First-Order Differential Equations

2.1 Linear Differential Equations

Definition 2.1 (First-Order, Linear Equation). An equation of the form

$$\frac{dy}{dt} + p(t)y(t) = q(t)$$

is said to be a linear, differential equation of the first order. If $q(t) \equiv 0$, then we say that this is a linear, **homogeneous** equation.

Let us now consider the linear equation with **variable coefficients**

$$\frac{dy}{dt} + p(t)\,y = q(t).$$

The solution of this equation may be represented as

$$y(t) = C \exp\left(-\int^t p(t)\,dt\right) + \exp\left(-\int^t p(t)\,dt\right) \int_{t_0}^t q(s) \exp\left(\int^s p(t)\,dt\right) ds,$$

where $\int^t p(t)\,dt$ denotes any indefinite integral of $p(t)$. We can write a MAPLE **procedure** which solves linear equations based on this formula, namely

```
>   LINODE1 := proc(p,q,t,C)
>   description
>   "general solution of first-order linear ode"; C*exp(-int(p,t)) +
>   exp(-int(p,t))* int(q*exp(int(p,t)),t)
>   end proc;
```

$LINODE1 := \mathbf{proc}(p, q, t, C)$
$\quad \mathbf{description}\,\text{``general solution of first-order\ linear ode''};$
$\qquad C * \exp(-\mathrm{int}(p,\ t)) + \exp(-\mathrm{int}(p,\ t)) * \mathrm{int}(q * \exp(\mathrm{int}(p,\ t)),\ t)$
$\quad \mathbf{end\ proc}$

```
> with(DEtools):
```
First we recall how one plots the solution of linear odes with continuous nonhomogeneous term using DEplot

```
> ode1:=diff(y(x),x)+exp(-x)*y(x)-x;
```

$$ode1 := \left(\frac{\partial}{\partial x} y(x)\right) + e^{(-x)} y(x) - x$$

```
> DEplot(ode1,y(x),x=-1..5,[[y(0)=1],[y(0)=-2]],stepsize=0.01);
```

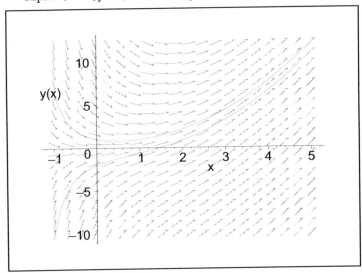

```
> DEplot(ode1,y(x),x=-1..5,[[y(0)=1],[y(0)=-2]],stepsize=0.03);
```

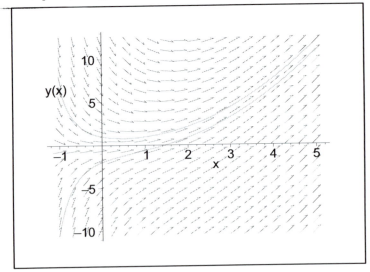

Now we consider how to handle differential equations with discontinuous **forcing** terms. First we need to know the correct syntax for introducing a discontinuous function by means of the Heaviside function, which is defined in MAPLE as

$$H(x) := \begin{cases} 1 & x > 0 \\ 0 & x < 0 \end{cases}.$$

An alternate way of introducing discontinuous functions is by using `type(x,numeric)` in imbedded `if` statements.

```
>   g:=(x)->
>   x*Heaviside(2-x)+(2-x)*Heaviside(x-2);
```

$$g := x \to x \, \text{Heaviside}(2 - x) + (2 - x) \, \text{Heaviside}(x - 2)$$

```
>   with(plots):
>   plot(g(x),x=-3..5);
```

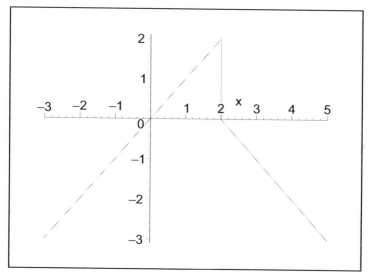

Let us check if MAPLE understands our definition of the function g.

```
>   with(plots):
```

Now let us introduce a differential equation having $g(x)$ as the nonhomogeneous term and see how MAPLE may be used to solve such an equation.

```
>   ode_2:=diff(y(x),x)+y(x)-g(x);
```

$$ode\_2 := \left(\frac{\partial}{\partial x} \, y(x)\right) + y(x) - x \, \text{Heaviside}(2 - x) - (2 - x) \, \text{Heaviside}(x - 2)$$

```
>   DEplot(diff(y(x),x)+y(x)-g(x),y(x),x=-1..3,y=-10..10,number=1,
>   [[y(0)=1],[y(0)=-2]]);
```

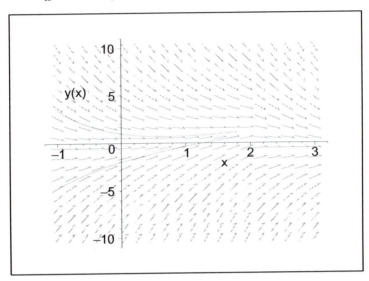

If we wish to obtain an analytical solution to this equation, we may use
our MAPLE program LINODE1, namely

```
>  y(t):=LINODE1(1,g(t),t,C);
```

$$Ce^{-t} + e^{-t}\left(-2\,Heaviside\,(t-2)\,(t-1)\,e^{t} + (t-1)\,e^{t} + 2\,Heaviside\,(t-2)\,e^{t}\right)$$

```
>  y1:=subs(C=1,y(t));y2:=subs(C=-2,y(t)):
```

$$y2 := -2\,e^{-t} + e^{-t}\left(-2\,Heaviside\,(t-2)\,(t-1)\,e^{t} + (t-1)\,e^{t} + 2\,Heaviside\,(t-2)\,e^{t}\right)$$

```
>  plot(y1,t=-1..5);
```

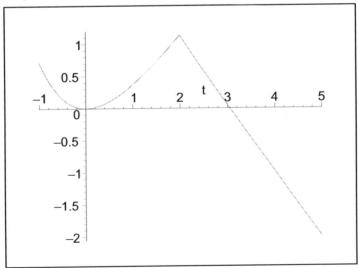

> `plot(y2,t=-1..5);`

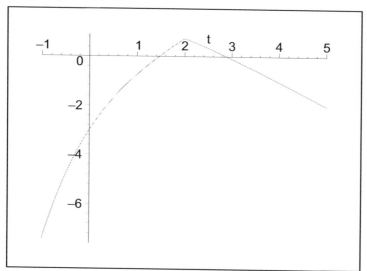

In general linear, ordinary differential equations are much easier to solve than **nonlinear** differential equations. A nonlinear differential equation which is reducible to a linear equation is the BERNOULLI equation. The Bernoulli differential equation has the form

$$\frac{dy}{dt} + p(t)y = q(t)y^n, n \neq 1.$$

The substitution $v = y^{1-n}$ reduces BERNOULLI's equation to a linear differential equation. Let us do this with MAPLE . First we input the Bernouilli equation

> `deq:= diff(y(x),x)+p(x)*y(x) =q(x)*y(x)^n;`

$$deq := (\frac{\partial}{\partial x}\, \mathrm{y}(x)) + \mathrm{p}(x)\,\mathrm{y}(x) = \mathrm{q}(x)\,\mathrm{y}(x)^n$$

Next we make the suggested substitution

> `deq2:= subs(y(x)=v(x)^(1/(1-n)),deq);`

$$(\frac{\partial}{\partial x}\, \mathrm{v}(x)^{(\frac{1}{1-n})}) + \mathrm{p}(x)\,\mathrm{v}(x)^{(\frac{1}{1-n})} = \mathrm{q}(x)\,(\mathrm{v}(x)^{(\frac{1}{1-n})})^n$$

> `deq2:=simplify(%);`

$$deq2 := \frac{\partial}{\partial x}\left((v(x))^{(1-n)^{-1}}\right) + p(x)\,(v(x))^{(1-n)^{-1}} = q(x)\left((v(x))^{(1-n)^{-1}}\right)^n$$

> `simplify(-deq2/v(x)^(-n/(-1+n)));`

$$deq2 := \frac{\partial}{\partial x}\left((v(x))^{(1-n)^{-1}}\right) + p(x)(v(x))^{(1-n)^{-1}} = q(x)\left((v(x))^{(1-n)^{-1}}\right)^{n}$$

The differential equation still looks complicated so let us make the coefficient of the derivative equal to 1.

```
> simplify((1-n)*%);
```

$$-\frac{d}{dx}v(x) + v(x)p(x)(-1+n) = (-1+n)(v(x))^{\frac{n}{-1+n}}q(x)\left((v(x))^{-(-1+n)^{-1}}\right)^{n}$$

We now recognize that the $v(x)$ terms on the right-hand-side cancel and we have obtained a first-order, linear equation.

Another non-linear equation we can use with linear methods is the Riccati equation

$$\frac{d\,y(x)}{dx} = q_1(x) + q_2(x)y(x) + q_3(x)y(x)^2.$$

It is convenient to input the right-hand-side as a function.

```
> with(DEtools):
> f:=(x,y) ->
> q[1](x)+q[2](x)*y(x)+q[3](x)*y(x)^2;
```

$$f := (x,\,y) \rightarrow q_1(x) + q_2(x)\,y(x) + q_3(x)\,y(x)^2$$

```
> deq:=diff(y(x),x)=f(x,y);
```

$$deq := \frac{\partial}{\partial x}\,y(x) = q_1(x) + q_2(x)\,y(x) + q_3(x)\,y(x)^2$$

As in the case of Bernoulli's equation, we seek to simplify the equation by making a suitable transformation of the unknown function.

```
> deq1:=subs(y(x)=y[1](x)+1/v(x),deq);
```

$$deq1 := \frac{\partial}{\partial x}\left(y_1(x) + \frac{1}{v(x)}\right) =$$
$$q_1(x) + q_2(x)\left(y_1(x) + \frac{1}{v(x)}\right) + q_3(x)\left(y_1(x) + \frac{1}{v(x)}\right)^2$$

```
> deq1:=simplify(deq1);
```

$$deq1 := \frac{d}{dx}y_1(x) - \frac{\frac{d}{dx}v(x)}{(v(x))^2} = q_1(x) + q_2(x)\left(y_1(x) + (v(x))^{-1}\right)$$
$$+ \frac{q_3(x)(y_1(x)v(x)+1)^2}{(v(x))^2}$$

```
> eq:=subs(y(x)=y[1](x),deq);
```

$$eq := \frac{\partial}{\partial x}\,y_1(x) = q_1(x) + q_2(x)\,y_1(x) + q_3(x)\,y_1(x)^2$$

```
> deq2:=subs(eq,deq1);
```

$$deq2 := \frac{(q_1(x) + q_2(x)\, y_1(x) + q_3(x)\, y_1(x)^2)\, \mathrm{v}(x)^2 - (\frac{\partial}{\partial x}\, \mathrm{v}(x))}{\mathrm{v}(x)^2} :$$

Notice: In order to simplify the presentation, we suppress Maple output in deq2 by ending the input command with a colon (:), rather than a semicolon(;)

```
>   deq3:=diff(v(x),x)=solve(deq2,diff(v(x),x));
```

$$deq3 := \frac{\partial}{\partial x}\, \mathrm{v}(x) = -q_2(x)\, \mathrm{v}(x) - 2\, q_3(x)\, y_1(x)\, \mathrm{v}(x) - q_3(x)$$

```
>   soln:=dsolve(deq3,v(x));
```

$$soln := \mathrm{v}(x) = \left(\int -q_3(x)\, e^{-\int -2\, q_3(x)y_1(x) - q_2(x)\, dx}\, dx + \_C1 \right) e^{\int -2\, q_3(x)y_1(x) - q_2(x)\, dx}$$

EXERCISES

Exercise 2.1. *Experiment with MAPLE to discover a method similar to that used above for finding* <u>*some*</u> *solution of*

$$\frac{dy}{dt} + ay = (t^2 + 3t + 2)e^t.$$

Generalize this to a right hand side of the form $(At^2 + Bt + C)e^t$.

Exercise 2.2. *Experiment with MAPLE to discover a method similar to that used in the previous problems for finding* <u>*some*</u> *solution of*

$$\frac{dy}{dt} + ay = (t^2 + 3t + 2)\sin t.$$

Exercise 2.3. *Write a MAPLE procedure that solves the* BERNOULLI *equation. Try it on the following examples.*

- $ty' + 3y = y^2$
- $y' + y = y^3$
- $y' - y = y^3$
- $t^2 y' + ty = y^2$

Exercise 2.4. *Write a MAPLE procedure for solving the following linear equation with discontinuous right-hand-side:*

$$\frac{dy}{dt} + ay = f(t), \text{where}$$

$$f(t) := \begin{cases} \alpha & \text{if } t \in [0,1] \\ \beta & \text{if } t \in [1,\infty) \end{cases}$$

Apply your procedure to the equation

$$\frac{dy}{dt} + y = f(t), \text{where}$$

$$f(t) := \begin{cases} 3 & \text{if } t \in [0,1] \\ 0 & \text{if } t \in [1,\infty) \end{cases}$$

```
plot your solution.
```

Exercise 2.5. *Write a MAPLE procedure for solving the following linear equation with discontinuous coefficient:*

$$\frac{dy}{dt} + p(t)y = f(t), \text{ where}$$

$$p(t) := \begin{cases} a_1 & \text{if } t \in [0,1] \\ a_2 & \text{if } t \in [1,\infty) \end{cases}$$

and f is continuous. Apply your procedure to the equation

$$\frac{dy}{dt} + p(t)y = 0, \text{ where}$$

$$p(t) := \begin{cases} 1 & \text{if } t \in [0,1] \\ -1 & \text{if } t \in [1,\infty) \end{cases}$$

plot *your solution.*

2.2   Project: Mixing Problems

2.2.1   Project1: One Tank

We consider first the situation where a tank contains initially a solution with the concentration C_0 [gm] [cm]$^{-3}$. To this solution the same solute but with a different concentration is added and stirred constantly. Call the concentration being added C_1 [gm] [cm]$^{-3}$, and the volumetric rate of solution being added r_i [cm]^3 [sec]$^{-1}$. Suppose at the same time the well-mixed fluid is drawn off at the volumetric rate of r_o [cm]^3 [sec]$^{-1}$. The problem is: what is the concentration of the solute in the tank as a function of the time t. We list below two MAPLE procedures which will be of help in analyzing this problem. Hint: Consider the three cases

- before overflow, for $r_i > r_o$

- after overflow

- before emptying, for $r_i < r_o$

Let us input the differential equation for a mixing problem in a partially filled tank. Suppose at time t the amount of the solute dissolved in the fluid in the tank is $q(t)$. If the initial volume of the tank is V_0, then after t seconds the volume is $V(t) := V_0 + (r_i - r_o)t$. Solute is flowing in at $r_i C_1$; whereas solute flows out at $r_o \frac{q(t)}{V_0 + (r_i - r_o)t}$. Hence, the differential equation for the amount of dissolved material is

$$\frac{dq(t)}{dt} = r_i C_i - \frac{r_o q(t)}{V_0 + (r_i - r_o)t}.$$

This solution to this equation may found by using `dsolve`. Let us first input the differential equation.

```
> eq1:=diff(q(t),t)-r_i*C_i+q(t)/(V_0+(r_i-r_o)*t);
```

$$eq1 := \left(\tfrac{\partial}{\partial t}\, q(t)\right) - r\_i\, C\_i + \frac{q(t)}{V\_0 + (r\_i - r\_o)\, t}$$

```
> y:=dsolve(eq1,q(t));
```

$$y := q(t) = \left(V\_0 + t\, r\_i - t\, r\_o\right)^{\left(-\frac{1}{r\_i - r\_o}\right)}\, C1$$
$$+ \frac{r\_i\, C\_i\, (V\_0 + t\, r\_i - t\, r\_o)}{1 + r\_i - r\_o}$$

The amount of solute initially in the tank is the initial volume multiplied by the initial concentration. Therefore we determine the constant of integration so that $q(0) = C_{init} V_0$ is satisfied.

```
> solve(C_init*V_0-subs(t=0,rhs(y)),_C1);
```

$$\frac{V\_0\,(C\_init + C\_init\, r\_i - C\_init\, r\_o - r\_i\, C\_i)}{V\_0^{\left(-\frac{1}{r\_i - r\_o}\right)}\,(1 + r\_i - r\_o)}$$

We then substitute this constant into the general solution to the equation.

```
> simplify(subs(_C1=%,y));
```

$$q(t) = \frac{1}{1 + r\_i - r\_o}\left(-V\_0^{\frac{1 + r\_i - r\_o}{r\_i - r\_o}}\left((C\_i - C\_init)r\_i + C\_init\,(r\_o - 1)\right)(V\_0 + (r\_i - r\_o)t)^{-(r\_i - r\_o)^{-1}} + r\_i\, C\_i\,(t\, r\_i - t\, r\_o + V\_0)\right)$$

Let us use this solution to construct a MAPLE function which solves this problem for the various values of the parameters r_i, r_o, C_0, C_1, V_0:

```
> mix_1:=proc(r_i,r_o,V_0,C_init,C_i,t)
> description "This program solves the mixing problem where the
> initial concentration, the concentration of solute added, the
> rates of influx and the initial volume are given." ;
> ((V_0+t*r_i-t*r_o)^(-1/(r_i-r_o))*V_0^((1+r_i-r_o)/(r_i-r_o))*C_init+(
> V_0+t*r_i-t*r_o)^(-1/(r_i-r_o))*V_0^((1+r_i-r_o)/(r_i-r_o))*C_init*r_i
> -(V_0+t*r_i-t*r_o)^(-1/(r_i-r_o))*V_0^((1+r_i-r_o)/(r_i-r_o))*C_init*r
> _o-(V_0+t*r_i-t*r_o)^(-1/(r_i-r_o))*V_0^((1+r_i-r_o)/(r_i-r_o))*r_i*C_
> i+r_i*C_i*V_0+r_i^2*C_i*t-r_i*C_i*t*r_o)/(1+r_i-r_o)
> end proc;
```

mix _1 := **proc**(r_i, r_o, V_0, C_init, C_i, t)

description "This program solves the mixing problem where\nthe initial co\ncentration, the concentration of solute added, the rates of influx and the i\nitial volume are given.";

$$((V\_0 + t * r\_i - t * r\_o)^{(-1/(r\_i - r\_o))} * V\_0^{((1+r\_i-r\_o)/(r\_i-r\_o))} * C\_init$$
$$+(V\_0 + t * r\_i - t * r\_o)^{(-1/(r\_i - r\_o))} * V\_0^{((1+r\_i-r\_o)/(r\_i-r\_o))} * C\_init$$
$$* r\_i - (V\_0 + t * r\_i - t * r\_o)^{(-1/(r\_i - r\_o))} * V\_0^{((1+r\_i-r\_o)/(r\_i-r\_o))} *$$
$$C\_init * r\_o-$$
$$(V\_0 + t * r\_i - t * r\_o)^{(-1/(r\_i - r\_o))} * V\_0^{((1+r\_i-r\_o)/(r\_i-r\_o))} * r\_i * C\_i$$
$$+ r\_i * C\_i * V\_0 + r\_i^2 * C\_i * t - r\_i * C\_i * t * r\_o)/(1 + r\_i - r\_o)$$

end proc

We now use this MAPLE function to solve the problem for an inflow rate of $r_i = 5$, an out flow rate of $r_o = 4$, a tank originally holding $V_0 = 100$ gallons of pure water, i.e. $C_0 = 0$, and with the solute flowing in having a concentration $C_1 = 1$. The solution is then given by

```
>   mix_1(5,4,100,0,1,t);
```

$$-25000\,\frac{1}{100+t} + 250 + \frac{5}{2}t$$

We may plot this solution as

```
>   plot(%,t=0..10,axes=BOXED);
```

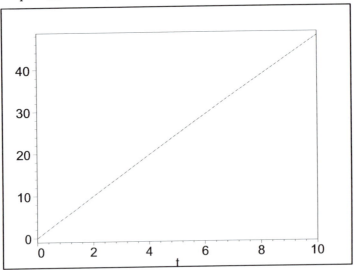

If this solution were to be valid for all time then the amount of material in the tank at $t = \infty$ would be

```
>   limit(%%,t=infinity);
```

$$\infty$$

Clearly this is not possible in a finite tank. Suppose the tank can only hold 110 gallons. Then overflow occurs in at 10 seconds and another differential equation governs the system. The amount of material in the tank at the end of 10 seconds is

> `limit(%%,t=10);`

$$\frac{525}{11}$$

The differential equation which governs the system now must have inflow equal to outflow. The volume of solute in the tank remains fixed at 110 gallons. The flow into the tank remains the same. If we assume the mixture is thoroughly stirred, then the amount which flows out is $\frac{q(t)}{V_1}r_i$. Recall the outflow now must equal the inflow. We solve the new differential equation

$$\frac{d\,q(t)}{dt} = C_i r_o - r_o \frac{q(t)}{V_0},$$

to get

> `deq2:=diff(q(t),t)+q(t)/V_0*r_o-C_ir_o;`

$$deq2 := (\tfrac{\partial}{\partial t}\,q(t)) + \frac{q(t)\,r\_o}{V\_0} - C\_ir\_o$$

> `dsolve({deq2,q(t_1)=q_1},q(t));`

$$q(t) = C\_i\,V\_0 - (C\_i\,V\_0 - q\_1)e^{-\frac{tr\_o}{V\_0}}\left(e^{-\frac{r\_ot\_1}{V\_0}}\right)^{-1}$$

> `y:=dsolve({eq1,q(0)=V_0*c_init},q(t));`

$$y := q(t) = \Big(\frac{(tr\_i - tr\_o + V\_0)^{1+(r\_i-r\_o)^{-1}}\,r\_i\,C\_i}{1+r\_i - r\_o}$$

$$-\frac{1}{1+r\_i-r\_o}(C\_i\,V\_0{}^{\frac{1+r\_i-r\_o}{r\_i-r\_o}}\,V\_0{}^{-(r\_i-r\_o)^{-1}}\,r\_i - V\_0\,c\_init\,r\_i +$$

$$V\_0\,c\_init\,r\_o - V\_0\,c\_init)(V\_0{}^{-(r\_i-r\_o)^{-1}})^{-1})(V\_0 + (r\_i - r\_o)t)^{-(r\_i-r\_o)^{-1}},$$

which can be simplified in the form :

$$q(t) = \frac{1}{1+r\_i - r\_o}(-V\_0{}^{\frac{1+r\_i-r\_o}{r\_i-r\_o}}((C\_i - c\_init)r\_i + c\_init\,(r\_o - 1))$$

$$* \ (V\_0 + (r\_i - r\_o)t)^{-(r\_i-r\_o)^{-1}} + r\_i\,C\_i\,(tr\_i - tr\_o + V\_0))$$

We now use the solution **deq2** to write a program which solves the second mixing problem.

```
> mix_2:=proc(r_o,V_0,C_i,q_1, t_1,t)
> description "This program computes the solute in the tank when the
> flow rate in equals the flow rate out" ;
> C_i*r_o*V_0/r_o+exp(-1/V_0*r_o*t)*(-C_i*r_o*V_0/r_o+q_1)/
> exp(-1/V_0*r_o*t_1)
> end proc;
```

$mix\_2 := \mathbf{proc}(r\_o, \ V\_0, \ C\_i, \ q\_1, \ t\_1, \ t)$

description "This program computes the solute in the tank when the flow rate in equals the flow rate out";

$C\_i * V\_0 + \exp(-r\_o * t/V\_0) * (-C\_i * V\_0 + q\_1)/\exp(-r\_o * t\_1/V\_0)$

end proc

Let us now compute the amount of material in the tank. Recall that the initial time is now 10 seconds and the initial amount of material was found to be $\frac{525}{11}$.

```
> simplify(mix_2(5,110,1,525/11,10,t));
```

$$110 - \frac{685}{11} e^{(-1/22\,t+5/11)}$$

Let us check the initial condition by substituting in $t = 10$

```
> subs(t=10,%);
```

$$110 - \frac{685}{11} e^0$$

```
> simplify(%);
```

$$\frac{525}{11}$$

We now combine these two MAPLE functions to obtain one that solves the composite system.

```
> mix :=proc(ri,ro,V_0,V_1,C_0,C_1,t) local
> t_m ; t_m:=(V_1-V_0)/(ri -ro); if  t<=t_m then
> mix1(ri,ro,V_0,C_init,C_i,t) else mix2(ri,V_1,C_1,
> evalf(mix1(ri,ro,V_0,C_0,C_1,t_m)),t_m,t)
> fi; end;
```

$mix := \mathbf{proc}(ri, \ ro, \ V\_0, \ V\_1, \ C\_0, \ C\_1, \ t)$

    local t_m;

      $t\_m := (V\_1 - V\_0)/(ri - ro)$;

      if $t \le t\_m$ **then** $mix1(ri, \ ro, \ V\_0, \ C\_0, \ C\_1, \ t)$

      else $mix2(ri, \ V\_1, \ C\_1, \ evalf(mix1(ri, \ ro, \ V\_0, \ C\_0, \ C\_1, \ t\_m)),$

      $t\_m, \ t)$

      fi

    end

Let us compute the solution at a sequence of times which extends past the time of overflow. We make the solutions at these times $t = k, k = 1, \ldots 12$ into a list of coordinate points

```
>   L:=[seq([k,evalf(mix(5,4,100,110,0,1,k))],k=1
>   ..12)];
```

$L := [[1, 4.975247525], [2, 9.901960784], [3, 14.78155340], [4, 19.61538462],$
$[5, 24.40476190], [6, 29.15094340], [7, 33.85514019], [8, 38.51851852],$
$[9, 43.14220183], [10, 47.72727273], [11, 50.49448366], [12, 53.13872812]]$

```
>   with(plots):
```
We now plot these points using `display`
```
>   g1:=plot(mix1(5,4,100,0,1,t),t=0..10):
>   :
>   g2:=plot(mix2(5,110,1,525/11,10,t),t=10..20)
>   display(g1,g2,axes=BOXED);
```

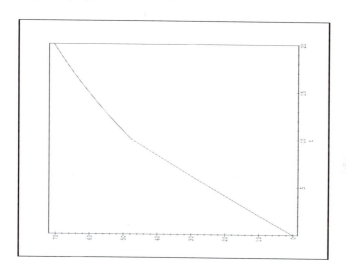

2.2.2 Project2: Two Tanks

We now consider a two tank problem. We will assume that there are two 100 liter tanks, one of which we will refer to as **A** and the other as **B**. In the tank **A** there are initially 50 liters of liquid of type α and 50 `liters` of liquid of type β. In tank **B** there are initially 100 `liters` of liquid of type α. Pure α fluid is pumped into tank **A** at the rate of 5 `[l][sec]`$^{-1}$. A well-stirred mixture is then drained off into tank **B** at the rate of 5 `[l][sec]`$^{-1}$. Tank **B** also has a drain pipe removing a well-stirred mixture at 5 `[l][sec]`$^{-1}$. We want to find the amount of β fluid in tank **B**. Let us denote the amount of the β liquid in tanks **A** and **B** by $c_A(t)$ and $c_B(t)$, respectively. First let us consider tank **A**. The change in liquid of type β in tank **A** is given by

$$\frac{d\,c_A(t)}{dt} = -(t_1/100)\,5\,[1]\,[\mathrm{sec}]^{-1}.$$

Whereas the change in liquid of type β in tank **B** is given by

$$\frac{d\,c_B(t)}{dt} = \frac{5}{100}\,(c_A - c_B)\,[l][sec]^{-1}$$

Write MAPLE procedure for solving the two-tank problem.

2.3   Separable Differential Equations

One of the simplest classes of differential equations, of first order, are the separable differential equations. These are differential equations that can be manipulated into the form

$$p(x)dx + q(y)dy = 0,$$

or

$$dy/dx = -\frac{p(x)}{q(y)}.$$

Once the equation is in this form, it may be solved directly by integration. We assume that the coefficients $p(x)$ and $q(y)$ are continuous, and seek a solution which passes through the initial point (x_0, y_0). The solution is then obtained in implicit form as

$$A(x) + B(y) = 0,$$

where

$$A(x) = \int_{x_0}^{x} p(x)dx, \quad B(y) = \int_{y_0}^{y} q(y)dy.$$

For example, suppose we wished to solve the initial value problem

$$3x^2y^2dx + (x^2 + 1)dy = 0,$$

$$y(0) = 1,$$

then we **separate** the equation into the form

$$\frac{3x^2dx}{x^2 + 1} + \frac{dy}{y^2} = 0,$$

or
$$3dx - \frac{3dx}{x^2 + 1} + \frac{dy}{y^2} = 0,$$

and integrate (for $y \neq 0$)

$$3 \int_0^x dx - 3 \int_0^x \frac{dx}{x^2 + 1} + \int_1 y \frac{dy}{y^2} = 0.$$

This becomes
$$3x - 3 \operatorname{atan} x - y^{-1} + 1 = 0,$$

or
$$y = (3(x - \operatorname{atan} x) + 1)^{-1}.$$

In order to satisfy the initial condition $y(0) = 1$, we choose the constant of integration to be $c = 1$. We note the practical problem of solving a separable differential is to discover a **separated** form of the equation. After this the remaining problem is just a matter of doing an integration. Once we have a separated form we could write a simple program to do the integration and solve for the dependent variable; however, MAPLE already has the capability of doing this separation as a part of the DEtools package. We will, therefore defer to MAPLE and load DEtools.

> with(DEtools):

As an example we enter the separable differential equation

> ode_1 := t^3*(z(t)+1) +z(t)^2*(t-1)^2*diff(z(t),t) = 0;

$$ode\_1 := t^3 (z(t) + 1) + z(t)^2 (t - 1)^2 \left(\tfrac{\partial}{\partial t} z(t)\right) = 0$$

The command separablesol actually does the separation and gives us the general solution to the equation.

> separablesol(ode_1, z(t));

$$\{\tfrac{1}{2} z(t)^2 - z(t) + \ln(z(t) + 1) + \tfrac{1}{2} t^2 + 2t - \frac{1}{t - 1} + 3\ln(t - 1) = \_C_1\}$$

Hence, separablesol produces a solution with an arbitrary constant. In order for the solution to pass through an initial point we need to choose the constant appropriately. We test separablesol on another equation.

> ode_2:=diff(y(t),t)-t*y(t)^3*(1-t)^(-1/2)=0;

$$ode\_2 := \left(\tfrac{\partial}{\partial t} y(t)\right) - \frac{t\, y(t)^3}{\sqrt{1 - t}} = 0$$

> sol:=separablesol(ode_2, y(t));

$$sol := \{-\frac{1}{2} \frac{1}{y(t)^2} - \frac{2}{3}(1 - t)^{(3/2)} + 2\sqrt{1 - t} = \_C_1\}$$

> subs(y(0) = 1, subs(t = 0, op(1, sol)));

$$1 = -3 \left(\sqrt{24 - 18 \_C1}\right)^{-1}$$

Let us now choose the constant $\_C_1$ so that the solution passes through the initial point $(\vec{x}_0, y_0) = (0, 1)$. We first substitute the initial point into the solution and then solve for $\_C_1$. In order to pick out the right hand side of the line involving `sol1` we use `op(1,sol)`. Experiment with this command!

```
>    subs(y(0)=1,subs(t=0,op(1,sol)));
```

This results in an expression for $\_C_1$.

$$\frac{5}{6} = \_C_1$$

We are now in a position to write a program to solve an initial value problem for a separable equation. We call this program `separable_initial`. The program calls up `separablesol` to do the separation and then solves for the constant $\_C_1$.

```
>    separable_initial:=proc(de,t0,y0) local
>    sols : sols:=separablesol(de,y(t));
>    subs(_C[1]=op(1,subs(y(0)=y0,subs(t=t0,op(1,sols)))) )
>    ,sols) end;
```

$separable\_initial := \mathbf{proc}(de,\ t0,\ y0)$
$\quad \mathbf{local}\ sols;$
$\qquad sols := separablesol(de,\ \mathrm{y}(t))\ ;$
$\qquad subs(\_C_1 = op(1,\ subs(\mathrm{y}(0) = y0,\ subs(t = t0,\ op(1,\ sols)))),\ sols)$
$\quad \mathbf{end}$

We now test this program on a previous, initial value problem.

```
>    separable_initial(ode_2,0,1);
```

$$\{-\frac{1}{2}\frac{1}{\mathrm{y}(t)^2} - \frac{2}{3}(1-t)^{(3/2)} + 2\sqrt{1-t} = \frac{5}{6}\}$$

We test the program on a new initial value problem.

```
>    ode_3:=diff(y(t),t)-3*y(t)^3-t*y(t)^3;
```

$$ode\_3 := (\tfrac{\partial}{\partial t}\,\mathrm{y}(t)) - 3\,\mathrm{y}(t)^3 - t\,\mathrm{y}(t)^3$$

```
>    sol_3:=separablesol( ode_3, y(t) );
```

$$sol\_3 := \{-\frac{1}{2}\frac{1}{\mathrm{y}(t)^2} - 3\,t - \frac{1}{2}t^2 = \_C_1\}$$

```
>    separable_initial(ode_3,0,1);
```

$$\{-\frac{1}{2}\frac{1}{\mathrm{y}(t)^2} - 3\,t - \frac{1}{2}t^2 = \frac{-1}{2}\}$$

2.4 Exact Equations

The differential equation

$$p(t, y)dt + q(t, y)dy = 0$$

is said to be **exact** in Γ if there exists a function $\varphi(t, y)$ with continuous first partial derivatives such that

$$\frac{\partial \varphi}{\partial t} = p, \quad \text{and} \quad \frac{\partial \varphi}{\partial y} = q.$$

Theorem 2.7. Let $p(t, y)$ and $q(t, y)$ be continuous in the rectangle Γ. Then the differential equation is exact there.

$$p(t, y)dt + q(t, y)dy = 0$$

is exact in Γ if and only if

$$\frac{\partial p}{\partial y} = \frac{\partial q}{\partial t}.$$

We introduce a maple function which solves an exact equation of the form

$$p(x, y)\,dx \; + \; q(x, y)\,dy \; = \; 0.$$

```
>   exact1:=(p,q,x,y,c)->int(p,x)+int(q-int(diff(
>   p,y),x),y)+c;
```

$$exact1 := (p, q, x, y, c) \rightarrow \int p\,dx + \int q - \int \mathrm{diff}(p, y)\,dx\,dy + c$$

We test this MAPLE function on the equation

$$(y\cos(x) + 2x\exp y))\,dx \; + \; \left(\sin(x) + x^2\exp y + 2\right)\,dy \; = \; 0$$

```
>   exact1(y*cos(x)+2*x*exp(y),sin(x)+x^2*exp(y)+
>   2,x,y,c);
```

$$\sin(x)\,y + x^2\,e^y + 2\,y + c$$

As another example we try $(4 - y)/x^2\,dx \; + \; (y^2 - 5*x)/(x*y^2)\,dy \; = \; 0$

```
>   exact1((4-y)/x^2,(y^2-5*x)/(x*y^2),x,y,c);
```

$$-\frac{4 - y}{x} + \frac{y + 5\dfrac{x}{y}}{x} - \frac{y}{x} + c$$

```
>  simplify(%);
```

$$\frac{-4\,y + y^2 + 5\,x + c\,x\,y}{x\,y}$$

We now write a procedure which checks first whether the equation is exact, and if so, then generates the solution.

```
>  exact_equ :=proc(p,q)
>  if(diff(p,y)-diff(q,x)=0)
>  then
>  int(p,x)+int(q-int(diff(p,y),x),y)=c
>  else 'not exact, method non-applicable' fi ;
>  end;
```

$exact\_equ := \mathbf{proc}(p, q)$

    $\mathbf{if}\operatorname{diff}(p, y) - \operatorname{diff}(q, x) = 0 \,\mathbf{then}\operatorname{int}(p, x) + \operatorname{int}(q - \operatorname{int}(\operatorname{diff}(p, y), x), y) = c$

    $\mathbf{else}\,`not\ exact,\ method\ non\text{-}applicable$'

    $\mathbf{fi}$

  $\mathbf{end}$

We test this procedure on the equation

$$\frac{x\,dx}{(x^2 + y^2)^{(3/2)}} + \frac{y\,dy}{(x^2 + y^2)^{(3/2)}} = 0.$$

```
>  exact_equ(x/(x^2+y^2)^(3/2),y/(x^2+y^2)^(3/2)
>  );
```

$$-\frac{1}{\sqrt{x^2 + y^2}} = c$$

The equation

$$\left(xy^2 + x^2 * y\right)\,dx + (x + y)\,dy = 0$$

is not exact. Let us see what our procedure does.

```
>  exact_equ(x*y^2+x^2*y,x+y);
```

        not exact, method non-applicable

```
>  Exact_equ :=proc(p,q)
>  if(diff(p,y)-diff(q,x)=0)
>  then exact1(p,q,x,y,c)
>  else 'not
>  exact, method non-applicable' fi ;
>  end;
```

> *Exact_equ* := **proc**(*p*, *q*)
>> **if** diff(*p*, *y*) − diff(*q*, *x*) = 0 **then** exact1(*p*, *q*, *x*, *y*, *c*)
>> **else** '*not exact, method non-applicable*'
>> **fi**
> **end**

```
> Exact_equ(x/(x^2+y^2)^(3/2),y/(x^2+y^2)^(3/2)
> );
```

$$-\frac{1}{\sqrt{x^2 + y^2}} + c$$

In general, partial differential equations are much more difficult to solve than an ordinary differential equation. However, we need only find *one* of solution of the integrating factor equation, and a special form or type might be easy to find. We consider special cases such as integrating factors which are functions of t or y alone. Two MAPLE procedures to test when this is the case are given below. Can you justify with paper and pencil calculation why these procedures work? In this session we seek some integrating factors which have a simple form. If μ is an integrating factor for

$$p\,dx \,+\, q\,dy \,=\, 0$$

then

$$\frac{\partial \mu p}{\partial y} \;=\; \frac{\partial \mu q}{\partial x}.$$

Hence, if the integrating factor only depends on y then $\left(\frac{\partial q}{\partial x} - \frac{\partial p}{\partial y}\right)/p$ is just a function of y. On the other hand, if the integrating factor only depends on x then $\left(\frac{\partial p}{\partial x} - \frac{\partial q}{\partial x}\right)/q$ is just a function of x. With this in mind let us introduce the MAPLE functions A and B.

```
> A:=(p,q)->(diff(q,x)-diff(p,y))/p;
```

$$A := (p, \, q) \to \frac{\left(\frac{\partial}{\partial x}\, q\right) - \left(\frac{\partial}{\partial y}\, p\right)}{p}$$

```
> B:=(p,q)->(diff(p,y)-diff(q,x))/q;
```

$$B := (p, \, q) \to \frac{\left(\frac{\partial}{\partial y}\, p\right) - \left(\frac{\partial}{\partial x}\, q\right)}{q}$$

```
> B(3*x*y+y^2,x^2+x*y);
```

$$\frac{x + y}{x^2 + x\,y}$$

```
>  simplify(%);
```

$$\frac{1}{x}$$

Hence we see that an integrating factor exists which is just a function of x, which we may solve for using the condition (2.4).

```
>  dsolve(diff(m(x),x)-1/x*m(x),m(x));
```

$$m(x) = x \ \_C1$$

We consider the equation $dx + (x/y - \sin(y)) = 0$. Let us see if it has an integrating factor which just depends on y.

```
>  B(1,x/y-sin(y));
```

$$-\frac{1}{y\left(\dfrac{x}{y} - \sin(y)\right)}$$

```
>  A(1,x/y-sin(y));
```

$$\frac{1}{y}$$

```
>  dsolve(diff(m(y),y)-1/y*m(y),m(y));
```

$$m(y) = y \ \_C1$$

We can combine these two tests to see if an integrating factor exists which depends either just on x or just on y.

```
>  decide:=proc(p,q)
>  if diff((diff(q,x)-diff(p,y))/p,x)=0
>  then
>  print('integrating factor is a function of y
>  alone')
>  elif diff((diff(q,x)-diff(p,y))/q,y)=0
>  then
>  print('integrating factor is a function of x
>  alone') fi;
>  end;
```

decide := **proc**(*p*, *q*)
　　　　if diff(((diff(*q*, *x*) − diff(*p*, *y*))/*p*, *x*) = 0 **then**
　　　　　　print('*integrating factor is a function of y alone*')
　　　　elif diff(((diff(*q*, *x*) − diff(*p*, *y*))/*q*, *y*) = 0 **then**
　　　　　　print('*integrating factor is a function of x alone*')
　　　fi
　　end

We test our procedure on the previous example.
```
>   decide(1,x/y-sin(y));
```

　　　　integrating factor is a function of y alone.

Chapter 3

Numerical Methods for First-Order Equations

3.1 Picard's Iteration Method and Semi-Batch Reactor

It is well-known that, in general, solutions of nonlinear equations cannot be found explicitly. Hence, implicit, or more precisely, numerical methods are of value for these problems. In this chapter, we discuss a numerical procedure known as the *method of successive approximations*, or *Picard's iteration method* for constructing approximate solutions of nonlinear equations. Picard's iteration method is one of the classical methods for constructing existence proofs of solutions of nonlinear equations, when such solutions exist.

3.2 An Existence and Uniqueness Theorem

We recall the following fundamental existence and uniqueness theorem for the initial-value problem.

Theorem 1. *Let the functions f and $\partial f / \partial y$ be continuous in some rectangle $\alpha < x < \beta, \gamma < y < \delta$ containing the point (x_0, y_0). Then, in some interval $x_0 - h < x, x_0 + h$ contained in $\alpha < x < \beta$, there is a unique solution $y = \varphi(x)$ of the initial value problem (IVP):*

$$\frac{dy}{dx} = f(x, y), \qquad y(x_0) = y_0.$$

We remark that the conditions stated in Theorem 1 are sufficient to guarantee the existence of a unique solution of the IVP in some interval containing the point x_0. However, the determination of the actual size of the interval may be difficult. Moreover, even if f does not satisfy the hypotheses of the theorem, it is still possible that a unique solution may exist.

3.3 Picard Iteration Method

The usual proof of the above theorem involves constructing a sequence of functions, known as the *Picard iterates* of the equation, and showing that they always converge on some interval containing x_0. The Picard iterates are inductively defined by the equations:

$$y_1(x) = y_0 + \int_{x_0}^x f(s, y_0)ds$$

$$y_2(x) = y_0 + \int_{x_0}^x f(s, y_1(s))ds$$

$$\vdots \quad \vdots$$

$$y_N(x) = y_0 + \int_{x_0}^x f(s, y_{N-1}(s))ds.$$

As an example, let us compute the first four Picard iterates of the IVP:

$$\frac{dy}{dx} = -3\frac{y}{x+1}, \quad y(0) = 1$$

First we write a Maple procedure for performing the Picard iterates. We mention, however, Maple has its own built-in procedure which we will demonstrate later. The procedure depends on the right-hand-side f of the differential equation $\frac{dy}{dx} = f$, the number of iterates N, the initial conditions x_0 and y_0 defined as $y(x_0) = y_0$, and the idependent variable x.

```
>    our_picard :=proc(f,x,x0,y0,N) local k;
>    description "This program computes an approximate solution to an
>    initial value problem for a first-order differential equation":
>    g(x,y):=f; y||0:=y0; for k from 0 to N do y||(k+1):=y0
>    +int(subs(y=y||k,g(x,y)),x); print(y||(k+1)); od;
>    end;
```

our_picard := **proc**(*f*, *x*, *x0*, *y0*, *N*)
  local *k*;
  description "This program computes an approximate solution to an initial value problem for a first-order differential equation";
    $g(x, y) := f$;
    $y||0 := y0$;
    for *k* **from** 0 **to** *N* **do**
      $y||(k + 1) := y0 + \text{int}(\text{subs}(y = y||k, g(x, y)), x)$; $\text{print}(y||(k + 1))$
    end do
  end proc

Let us try this on the example now. We shall see it works well.

```
> our_picard(-3*y/(x+1),x,0,1,4);
```

$$1 - 3\ln(x + 1)$$

$$1 - 3\ln(x + 1) + \frac{9}{2}\ln(x + 1)^2$$

$$1 - 3\ln(x + 1) + \frac{9}{2}\ln(x + 1)^2 - \frac{9}{2}\ln(x + 1)^3$$

$$1 - 3\ln(x + 1) + \frac{9}{2}\ln(x + 1)^2 - \frac{9}{2}\ln(x + 1)^3 + \frac{27}{8}\ln(x + 1)^4$$

$$1 - 3\ln(x + 1) + \frac{9}{2}\ln(x + 1)^2 - \frac{9}{2}\ln(x + 1)^3 + \frac{27}{8}\ln(x + 1)^4 - \frac{81}{40}\ln(x + 1)^5$$

To simplify the output, let us change the procedure to just print the last iterate. We call this our_last_picard.

```
> our_last_picard :=proc(f,x,x0,y0,N) local
> k; description "This program computes an approximate solution to
> an initial value problem for a first-order differential equation":
> g(x,y):=f; y||0:=y0; for k from 0 to N do y||(k+1):=y0
> +int(subs(y=y||k,g(x,y)),x); print(y||(N+1)); od;
> end;
```

our_last_picard := **proc**(*f*, *x*, *x0*, *y0*, *N*)
  local *k*;
  description "This program computes an approximate solution to an initial value problem for a first-order differential equation";
    $g(x, y) := f$;
    $y||0 := y0$;
    for *k* **from** 0 **to** *N* **do**
      $y||(k + 1) := y0 + \text{int}(\text{subs}(y = y||k, g(x, y)), x)$; $\text{print}(y||(N + 1))$
    end do
  end proc

The last example was a **linear** problem and the Picard algorithm should have worked well! Now let us try the Picard method on a **non-linear** problem.

We consider the differential equation

$$\frac{dy}{dx} = y^2, \text{ with the initial condition} y(0) = 1.$$

We will see that the expression becomes quite complicated after four iterations. Try the method on a still more complicated non-linear differential equation, say

$$\frac{dy}{dx} = sin(y), \text{ with the initial condition} y(0) = 1,$$

and Maple will be stumped. Actually the Picard method is really not a good method for solving non-linear equations. However, we can improve the method!

What makes the Picard method impractical, is that at each step the function to be integrated becomes more and more complicated. Eventually any computer algebra system will be stumped and can go no further if we use it to solve non-linear equations with the Picard algorithm. However, if we can insure that the functions are always simple to integrate the method becomes an effective one. We can do this by expanding the iterate at each step as a Taylor series, and then expanding the integrand also as a Taylor series. The following procedure does exactly this.

```
>  taylor_picard:=proc(f,x,x0,y0,N,M)
>  description "This program approximates the output generated at
>  each stage of the iteration process by a Taylor series.";
>  local k;g(x,y):=f;
>  y||0:=y0; for k from 0 to N do y||(k+1):=taylor(y0
>  +int(taylor(subs(y=y||k,g(x,y)),x=0,M),x),x=0,M); od;
>  print(y||(N+1));
>  end:
>  taylor_picard(x*y^2,x,0,1,4,20);
```

$$1 + \frac{1}{2}x^2 + \frac{1}{4}x^4 + \frac{1}{8}x^6 + \frac{1}{16}x^8 + \frac{1}{32}x^{10} + \frac{43}{2880}x^{12}$$
$$+ \frac{13}{1920}x^{14} + \frac{943}{322560}x^{16} + \frac{3497}{2903040}x^{18} + O(x^{20})$$

```
>  with(plots):
```

Let us test the method on the previous example. If you try this scheme on various nonlinear equations, you will notice that the procedure is very much faster than the previous Picard procedure. On the linear problem we do here there is no advantage. As the output is lengthy we suppress it.

```
>  y2:=taylor_picard(-3*y/(x+1),x,0,1,4,40):
>  convert(%,polynom):
>  g2:=plot(%,x=0..1,y=-2..2,style=POINT,symbol=CROSS):
>  g1:=plot(1-3*ln(x+1)+9/2*ln(x+1)^2-9/2*
>  ln(x+1)^3+27/8*ln(x+1)^4-81/40
>  *ln(x+1)^5,x=0..1,y=-2..2,style=POINT,symbol=CIRCLE):
```

```
>   with(DEtools):
>   dsolve({D(y)(x)=-3*y(x)/(x+1),y(0)=1},y(x));
```

$$y(x) = \frac{1}{(x+1)^3}$$

```
>   g3:=plot(1/(x+1)^3,x=0..1):
```

We now compare these solutions. With the `taylor-picard` we took many terms quickly, but still this was not so accurate. This is a problem with Taylor series in general. However if we compare with a nonlinear problem we have a real advantage as `my_picard` will usually crash!

```
>   display({g1,g2,g3});
```

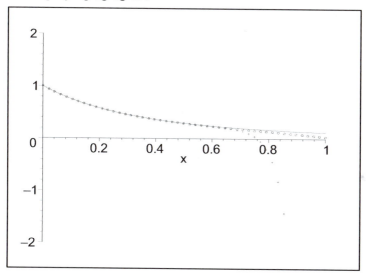

Now we consider the highly nonlinear differential equation

$$\frac{d\,y(x)}{dx} - \exp y(x) = 0,$$

again with the same initial conditions. Even though this equation is nonlinear, the solution is well known and Maple computes it easily. `our_picard` has real difficulties now. At the second term of iteration it obtains the answer in terms of the function $\mathrm{Ei}(1, y)$ and is able to go no further. It is unable to evaluate this function as it depends on the unknown $y(x)$. So the Picard algorithm is finished! However, `taylor_picard` works nicely and is accurate.

```
>   F:=our_picard(exp(y),x,0,1,2);
```

$$F := 1 - \mathrm{Ei}(1, -\frac{e^{(1+e\,x)}}{e})$$

```
>  y1:=taylor_picard(exp(y),x,0,0,4,40);
```

$$x + \frac{5}{36} x^7 + \frac{1187}{10080} x^8 + \frac{36403}{362880} x^9 + \frac{11513}{134400} x^{10} + \frac{162367}{2217600} x^{11}$$

$$+ \frac{1110281}{17740800} x^{12} + \frac{11890339}{222393600} x^{13} + \frac{1}{2} x^2 + \frac{1}{3} x^3 + \frac{1}{4} x^4 + \frac{1}{5} x^5 + \frac{119}{720} x^6$$

$$+ \frac{17070504284086081491969240467466351877}{17713967942376248450857116578611200000000} x^{36}$$

$$+ \frac{960741048420941674125256538032033233563}{12073467623882758812557876825948160000000} x^{37}$$

$$+ \frac{105942780734978325791986909678565875 9067}{16142673378598799745679235311730688000000 00} x^{38}$$

$$+ \frac{31508152756751482334106035289425327657 0747}{582796630891355524532579478282939924480000 0000} x^{39}$$

$$+ \frac{3262010872084072960738867}{640521732377550127104000000} x^{27}$$

$$+ \frac{64882230826998041987674391 3}{15244417230585693025075200000 0} x^{28}$$

$$+ \frac{314043031750231178803114985 03}{884176199373970195454361600000 0} x^{29}$$

$$+ \frac{1137839045091276536762870342 9}{3844244345104218241105920000000} x^{30}$$

$$+ \frac{96446870539523372967837941659 7}{39156374543704394370121728000000 0} x^{31}$$

$$+ \frac{3168347348498678958886831850068 9}{15478284525511384127483412480000000} x^{32}$$

$$+ \frac{3352646828428665917385842977330 13}{1973481277002701476254135091200000 00} x^{33}$$

$$+ \frac{2078579685994003748804429897113843 87}{1476163995198020704238093048217600000 00} x^{34}$$

$$+ \frac{3630115950445907207749}{5965850016665763840000 00} x^{26} + \frac{739771703127828419}{51090942171709440000} x^{21}$$

$$+ \frac{623961831475334959}{51090942171709440000} x^{22} + \frac{269105906104374107}{2616600884502528000 0} x^{23}$$

$$+ \frac{1788381838066794315739}{20681613391107981312000 0} x^{24} + \frac{2252002871970263267234 3}{31022420086661971968000 00} x^{25}$$

$$+ \frac{250035316477}{10461394944000} x^{18} + \frac{1232015888066479}{60822550204416000} x^{19} + \frac{10015172966221}{355687428096000} x^{17}$$

$$+ \frac{3978258079}{87178291200} x^{14} + \frac{3391525619}{87178291200} x^{15}$$

$$+ \frac{230995224791}{6974263296000} x^{16} + \frac{10424728164220661}{608225502044160000} x^{20}$$

$$+ \frac{547473937543252165076188712779276 93}{4696885439266429513484841517056000 0000} x^{35}$$

```
>   with(plots):
>   g1:=plot(y1,x=0..2,y=0..10,style=POINT,symbol=CROSS)
>   :
>   with(DEtools):
>   dsolve({diff(y(x),x)-exp(y(x)),y(0)=0},y(x));
```

$$y(x) = \ln(-\frac{1}{x-1})$$

```
>   g2:=plot(ln(-1/(x-1)),x=0..2,y=0..10):
>   display({g1,g2});
```

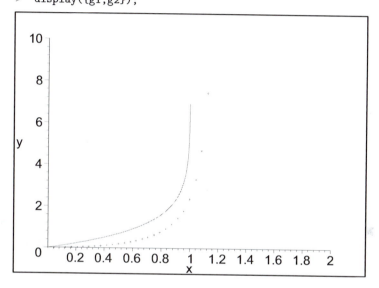

So `taylor_picard` works very well!

3.4 Computer Lab

Consider the semi-batch reactor shown in Figure 3.1. It is called a semi-batch reactor because there is flow into the reactor while no fluid is removed from the reactor.

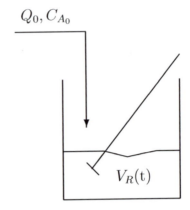

$$Q_0, C_{A_0}$$

$$V_R(t)$$

FIGURE 3.1. SEMI-BATCH REACTOR.

The following reaction occurs in the reactor:

$$A \xrightarrow{r} P$$

when $r = kC_A^2$. Initially, the reactor is filled to a volume V_0 with an inert liquid. At time equal to zero, a stream containing A at a concentration C_{A_0} is fed to the reactor at a flow rate Q_0. Performing an unsteady-state mole balance on A for this reactor in the following:

$$IN - CONSUMED + GENERATED = OUT + ACCUMULATION$$

$$Q_0 C_{A_0} - kC_A^2 V_R + 0 = 0 + \frac{d(C_A V_R)}{dt}.$$

Let n_A denote $C_A V_R$. Then this equation can be rewritten in the form

$$\frac{dn_A}{dt} = Q_0 C_{A_0} - \frac{kn_A^2}{V_R}.$$

Since liquid is added to the reactor, the volume of the reactor, V_R, will increase with time. Performing an overall mass balance on the reactor,

$$\frac{d}{dt}(\rho V_R) = Q_0 \rho$$

Then assuming ρ is a constant,

$$\frac{dV_R}{dt} = Q_0.$$

Integrating and using the initial condition yields the following:

$$V_R(t) = Q_0 t + V_0.$$

Using this result in the unsteady-state mole balance on A for the reactor results in the IVP:

$$\frac{dn_A}{dt} = Q_0 C_{A_0} - \frac{k n_A^2}{Q_0 t + V_0} \qquad (E)$$

where $n_A = 0$ at $t = 0$.

- Use the Picard iteration method to compute the first three iterates of the IVP.

- Write a modified Picard-Taylor iteration code to compute the first three iterates of the IVP and compare these with the Picard iteration results.

- Assume the following values for the parameters of the problem: $C_{A_0} = 1.0$ gmoles/liter, $k = 0.1$ liters/gmoles-sec, $Q_0 = 10.0$ liters/sec and $V_0 = 50$ liters. Plot the iterates for t from 0 to 100 seconds.

- The ODE (E) is a special form of the so-called *Riccati Equations*:

$$\frac{dy}{dt} = q_1(t) + q_2(t)y + q_3(t)y^2.$$

Hence, if we write $n_A(t) = n_p(t) + \frac{1}{v(t)}$ for any solution $n_p(t)$ of (E), then $v(t)$ satisfies a first-order linear ODE. Assume that $n_p(t)$ is known and find $v(t)$ in terms of $n_p(t)$

- Let $n_{A(N)}(t)$ and $n_{A(N-1)}(t)$ denote the Nth and $(N-1)$th iterates of $n_A(t)$, respectively. Find the number N so that

$$|n_{AN}(100) - n_{A(N-1)}(100)| \le 10^{-1}$$

3.5 Numerical Procedures and Fermentation Kinetics

In *real-world* applications of differential equations, solutions are often approximated since closed form solutions frequently cannot be found. Consequently, the role of approximate solutions is very important in practical problems. Here we introduce three basic numerical procedures for treating initial-value problems for first-order differential equations of the form:

IVP $\qquad\qquad \dfrac{dy}{dx} = f(x, y), \quad y(x_0) = y_0.$

The three fundamental procedures are *Euler's method, the improved Euler's method,*[1] and *Runge-Kutta's method*. By a numerical procedure for

---

[1] The improved Euler method is also known as the *Heun's method*

solving the IVP, we mean an algorithm for calculating approximate values $y_0, y_1, y_2, ..., y_n, ...$ of the solution y at a set of points $x_0 < x_1 < x_2 < \cdots < x_n < ...$. In the subsequent discussion we will always use a uniform spacing or step size h for the x-variable. Thus we have $x_1 = x_0 + h, x_2 = x_1 + h = x_0 + 2h$, and in general $x_n = x_0 + nh$.

3.6   The Euler Method

We begin with the Euler method, which is the simplest one of the three. Although it is not as accurate as the other two, it plays a fundamental role in the development of numerical algorithms for solving initial-value problems. The algorithm can be formulated as follows: With y_0 given from the initial condition, define the sequence $\{y_n\}$ by the recursion relation:

$$y_{n+1} = y_n + hf(x_n, y_n), \quad n = 0, 1, \cdots.$$

The y_n's are then used to approximate the exact solution $y(x)$ at $x = x_n$. There are many ways to derive or explain this algorithm. For instance, the Taylor series expansion for $y(x_{n+1})$ about x_n yields

$$y(x_{n+1}) = y(x_n) + h\frac{dy}{dx}(x_n) + \tau(x, h),$$

where $\tau(x, h)$ is referred to as the *local truncation error*. If $y(x)$ is twice continuously differentiable and has a bounded $d^2y/dx^2(x)$ in the region under consideration, then we have

$$\tau = O(h^2).$$

We notice that in the Taylor series expansion, we may write $dy(x_n)/dx = f(x_n, y(x_n))$ from the differential equation, and this leads to the Euler algorithm by neglecting the truncation term. We see that y_n will be a good approximation of $y(x_n)$ for h sufficiently small. [2] Indeed, under suitable assumptions of f, it can be shown that

$$|y(x_n) - y_n| = O(h^p) \quad \text{with} \quad p = 1.$$

The power of h indicates the order of convergence of the numerical procedure and hence the Euler method is a **first-order method**.

---

[2] We have adopted the Landau **order symbols** O and o which are defined as follows: *Given two functions $f(h)$ and $g(h)$, we write $f = O(g)$ as $h \to 0^+$, if $|f(h)/g(h)|$ is bounded as $h \to 0^+$, while we write $f = o(g)$ as $h \to 0^+$ if $f(h)/g(h) \to 0$ as $h \to 0^+$.*

3.7 Higher-Order Methods

The Euler method is not very practical, because of slow convergence; hence, we need **higher-order methods**. The *improved Euler's method* is a second-order method, and one of the most popular methods is a fourth-order *Runge-Kutta method*, which is a highly accurate numerical method often used by professional engineers. In what follows, we merely present these algorithms. We remark that the derivation of these algorithms can be obtained from the Taylor series expansions as was the case of the Euler method; however, details will be omitted

Improved Euler's Method

$$y_{n+1} = y_n + \frac{h}{2}(k_{n1} + k_{n2}), \quad n = 0, 1, \cdots,$$

where k_{n1} and k_{n2} are defined by

$$
\begin{aligned}
k_{n1} &= f(x_n, y_n), \\
k_{n2} &= f(x_n + h, y_n + hk_{n1}).
\end{aligned}
$$

Runge-Kutta's Method

$$y_{n+1} = y_n + \frac{h}{6}(k_{n1} + 2k_{n2} + 2k_{n3} + k_{n4}), \quad n = 0, 1, \cdots,$$

where k_{n1}, k_{n2}, k_{n3} and k_{n4} are defined by

$$
\begin{aligned}
k_{n1} &= f(x_n, y_n), \\
k_{n2} &= f(x_n + \frac{1}{2}h, y_n + \frac{1}{2}hk_{n1}), \\
k_{n3} &= f(x_n + \frac{1}{2}h, y_n + \frac{1}{2}hk_{n2}), \\
k_{n4} &= f(x_n + h, y_n + hk_{n3}).
\end{aligned}
$$

In view of its formulation, the improved Euler's method is sometime also referred to as the second-order Runge-Kutta method. It is not difficult to prove that the improved Euler's method and the Runge-Kutta method have local truncation errors of $O(h^3)$ and $O(h^5)$, respectively.

3.8 Maple Procedures

On Sun stations these three methods have been implemented in Maple procedures, which are contained in the ODE file. The ODE file must be read in each Maple session in order to perform the calculations. These procedures are called `firsteuler, impeuler,` and `rungekutta.` The arguments of all these procedures are `(f,init,h,N)`.Here f is the right side of the differential equation, `init`=$[x_0, y_0]$, h= step size, and N=number of iterations. However, these procedures are not available on the PC version of Maple. Hence, we have written our own procedures to do these. As an example, we consider the IVP:

$$\frac{dy}{dx} = x^2 + y, \quad y(0) = 1.4.$$

The following calculations are done by using the Maple procedures: First we write a procedure for the Euler algorithm. We call it `plain_euler` as the name euler is **protected** by Maple. The procedure depends on the right hand side $f(x, y)$, of the differential equation

$$\frac{d\,y(x)}{dx} = f(x, y),$$

the initial date $y_0 = y(x_0)$, the step size h, and the number of steps N.

```
>   plain_euler:=proc(f,x,x0,y0,h,N)
>   description "This is the simplest
>   numerical scheme for solving
>   inialvalue problem.";
>   g(x,y):=f:
>   x||0:=x0; y||0:=y0; for n from 0 to N do
>   x||(n+1):=evalf(x||n +h);
>   y||(n+1):=evalf(y||n + h*subs({x=x||n, y=y||n},g(x,y)));
>   print([x||(n+1),y||(n+1)]);
>   od;end;
```

Warning, 'n' is implicitly declared local to procedure
'plain_euler'

$plain\_euler := \textbf{proc}(f, x, x0, y0, h, N)$
local n;
description
"This is the simplest numerical scheme for solving inialvalue problem.";
$g(x, y) := f$;
$x||0 := x0$;
$y||0 := y0$;
for n **from** 0 **to** N **do**
$x||(n + 1) := \text{evalf}(x||n + h)$;
$y||(n + 1) := \text{evalf}(y||n + h * \text{subs}(\{x = x||n, y = y||n\}, g(x, y)))$;
$\text{print}([x||(n + 1), y||(n + 1)])$
end do
end proc

We test the procedure `plain_euler` on our problem with a step size $\frac{2}{10}$ and with 10 steps.

```
>   plain_euler(x^2+y,x,0,1.4,2/10,10);
```

$$[.2000000000, 1.680000000]$$
$$[.4000000000, 2.024000000]$$
$$[.6000000000, 2.460800000]$$
$$[.8000000000, 3.024960000]$$
$$[1.000000000, 3.757952000]$$
$$[1.200000000, 4.709542400]$$
$$[1.400000000, 5.939450880]$$
$$[1.600000000, 7.519341056]$$
$$[1.800000000, 9.535209268]$$
$$[2.000000000, 12.09025112]$$
$$[2.200000000, 15.30830134]$$

We shall see in what follows that the `plain_euler` answer at $x = 2.000000000$ of 12.09025112 is too small. Both the `improved_euler` and `runge_kutta` procedures lead to more accurate answers.

Next we input the procedure `improved_euler`.

```
>   improved_euler:=proc(f,x,x0,y0,h,N)
>   description "The improved euler method is
>   sometimes refered to as the
>   second-order Runge-Kutta method,
>   which leads to more acurate approximations.";
>   g(x,y):=f:
>   x||0:=x0; y||0:=y0;
>   k1||0:=subs({x=x0,y=y0},g(x,y));
>   k2||0:=subs({x=x0+h,y=y0+h*k1||0},g(x,y)); for n from 0 to N do
>   x||(n+1):=evalf(x||n +h);y||(n+1):=evalf(y||n+h/2*(k1||n +k2||n));
>   k1||(n+1):=subs({x=x||(n+1),y=y||(n+1)},g(x,y));
>   k2||(n+1):=subs({x=x||(n+1)+h,
>   y=y||(n+1)+h*k1||(n+1)},g(x,y));print([x||(n+1),y||(n+1)]); od;
>   end;
```

```
Warning, 'n' is implicitly declared local to procedure
'improved_euler'
```

improved_euler := **proc**(*f*, *x*, *x0*, *y0*, *h*, *N*)
local *n*;
description "The improved euler method is sometimes refered to as the second-order Runge-Kutta method,\nwhich leads to more acurate approximations.";

g(*x*, *y*) := *f* ;
x||0 := *x0* ;
y||0 := *y0* ;
k1||0 := subs({*x* = *x0*, *y* = *y0*}, g(*x*, *y*)) ;
k2||0 := subs({*x* = *x0* + *h*, *y* = *y0* + *h* ∗ *k1*||0}, g(*x*, *y*)) ;
for *n* **from** 0 **to** *N* **do**
    x||(*n* + 1) := evalf(*x*||*n* + *h*) ;
    y||(*n* + 1) := evalf(*y*||*n* + 1/2 ∗ *h* ∗ (*k1*||*n* + *k2*||*n*)) ;
    k1||(*n* + 1) := subs({*x* = *x*||(*n* + 1), *y* = *y*||(*n* + 1)}, g(*x*, *y*)) ;
    k2||(*n* + 1) := subs({*x* = *x*||(*n* + 1)
    + *h*, *y* = *y*||(*n* + 1) + *h* ∗ *k1*||(*n* + 1)},
    g(*x*, *y*)) ; print([*x*||(*n* + 1), *y*||(*n* + 1)])
end do
end proc

we next use improved euler on the problem, where the stepsize *h* is taken to be 0.2 and *N* = 10.

```
>   improved_euler(x^2+y,x,0,1.4,2/10,10);
```

$$[.2000000000, 1.712000000]$$
$$[.4000000000, 2.109440000]$$
$$[.6000000000, 2.628716800]$$
$$[.8000000000, 3.314234496]$$
$$[1.000000000, 4.220166085]$$
$$[1.200000000, 5.412602624]$$
$$[1.400000000, 6.972175201]$$
$$[1.600000000, 8.997253745]$$
$$[1.800000000, 11.60784957]$$
$$[2.000000000, 14.95037648]$$
$$[2.200000000, 19.20345931]$$

Let us use Maple to get the exact solution. We input the differential equation and the initial condition, and then dsolve.

```
>   with(DEtools):
>   deq1:=diff(u(x),x)=x^2+u(x);
```

$$deq1 := \frac{\partial}{\partial x}\, u(x) = x^2 + u(x)$$

```
> ic:=u(0)=1.4;
```

$$ic := u(0) = 1.4$$

```
> w:=dsolve({deq1,ic},u(x));
```

$$w := u(x) = -x^2 - 2\,x - 2 + \frac{17}{5}\,e^x$$

Let us check the error where we believe it will be, due to truncation build up, a maximum.

```
> subs(x=2,w);
```

$$u(2) = -10 + \frac{17}{5}\,e^2$$

The exact solution at 2.000000000, namely 15.12279074, compared with, the result we got from `improved_euler`, 14.95037648, is not bad. However, the Runge-Kutta algorithm will give a somewhat better numerical result at this point, as it should, namely 15.08480792.

```
> evalf(%);
```

$$u(2) = 15.12279074$$

Now let us write a **runge-kutta** procedure

```
> runge_kutta:=proc(f,x,x0,y0,h,N)
> description "This procedure is denoted by rkf45
> provided by Maple using a Fehlberg fourth-fifth order
> Runge-Kutta method.";
> g(x,y):=f: x||0:=x0; y||0:=y0; k1||0:=subs({x=x0,y=y0},g(x,y));
> k2||0:=subs({x=x0+h/2,y=y0+h/2*k1||0},g(x,y));
> k3||0:=subs({x=x0+h/2,y=y0+h/2*k2||0},g(x,y));
> k4||0:=subs({x=x0+h/2,y=y0+h/2*k3||0},g(x,y));
> for n from 0 to N do
> x||(n+1):=evalf(x||n+h);
> y||(n+1):=evalf(y||n+h/6*(k1||n+2*k2||n+2*k3||n+k4||n));
> k1||(n+1):=subs({x=x||(n+1),y=y||(n+1)},g(x,y));
> k2||(n+1):=subs({x=x||(n+1)+h/2,y=y||(n+1)+h/2*k1||(n+1)},g(x,y));
> k3||(n+1):=subs({x=x||(n+1)+h/2,y=y||(n+1)+h/2*k2||(n+1)},g(x,y));
> k4||(n+1):=subs({x=x||(n+1)+h,y=y||(n+1)+h*k3||(n+1)},g(x,y));
> print([x||(n+1),y||(n+1)]);
> od;
> end;
```

Warning, `n` is implicitly declared local to procedure `runge_kutta`

$runge\_kutta := \mathbf{proc}(f, x, x0, y0, h, N)$

local n;

description "This procedure is denoted by rkf45 \nprovided by Maple using a Fehlberg fourth-fifth order \nRunge-Kutta method.";

$g(x, y) := f$;

$x||0 := x0$;

$y||0 := y0$;

$k1||0 := \mathrm{subs}(\{x = x0, y = y0\}, \mathrm{g}(x, y))$;

$k2||0 := \mathrm{subs}(\{x = x0 + 1/2 * h, y = y0 + 1/2 * h * k1||0\}, \mathrm{g}(x, y))$;

$k3||0 := \mathrm{subs}(\{x = x0 + 1/2 * h, y = y0 + 1/2 * h * k2||0\}, \mathrm{g}(x, y))$;

$k4||0 := \mathrm{subs}(\{x = x0 + 1/2 * h, y = y0 + 1/2 * h * k3||0\}, \mathrm{g}(x, y))$;

for n **from** 0 **to** N **do**

   $x||(n+1) := \mathrm{evalf}(x||n + h)$;

   $y||(n+1) := \mathrm{evalf}(y||n + 1/6 * h * (k1||n + 2 * k2||n + 2 * k3||n + k4||n))$;

   $k1||(n+1) := \mathrm{subs}(\{x = x||(n+1), y = y||(n+1)\}, \mathrm{g}(x, y))$;

   $k2||(n+1) :=$
      $\mathrm{subs}(\{y = y||(n+1) + 1/2 * h * k1||(n+1), x = x||(n+1)$
      $+ 1/2 * h\}, \mathrm{g}(x, y))$;

   $k3||(n+1) :=$
      $\mathrm{subs}(\{y = y||(n+1) + 1/2 * h * k2||(n+1), x = x||(n+1)$
      $+ 1/2 * h\}, \mathrm{g}(x, y))$;

   $k4||(n+1) := \mathrm{subs}(\{y = y||(n+1) + h * k3||(n+1),$
   $x = x||(n+1) + h\}, \mathrm{g}(x, y))$;

   $\mathrm{print}([x||(n+1), y||(n+1)])$

end do

end proc

We test **runge_kutta** on the same problem.

```
>   runge_kutta(x^2+y,x,0,1.4,2/10,10);
```

$$[.2000000000, 1.706550000]$$

$$[.4000000000, 2.104602837]$$

$$[.6000000000, 2.625912572]$$

$$[.8000000000, 3.315480282]$$

$$[1.000000000, 4.228270283]$$

$$[1.200000000, 5.431415990]$$

$$[1.400000000, 7.006914157]$$

$$[1.600000000, 9.054915617]$$

$$[1.800000000, 11.69774460]$$

$$[2.000000000, 15.08480792]$$

$$[2.200000000, 19.39859106]$$

By a comparison of the three approximations, it is clear that the Runge-Kutta method gives the best approximation for the exact solution.

Now in order to illustrate the basic idea for writing procedures of this kind, we include here a typical procedure EULER(f,init,h,N) for the Euler method which is a simple procedure and works equally well as the one, firsteuler, contained in the ODE file.

```
> # This is the code for implementing the Euler method.
> EULER:=proc(f,init,h,N)
local i, X, Y, pts; g(x,y):=f: pts:=array(0..N); X:=init[1];
Y:=init[2]; pts[0]:=[X,Y]; for i to N do y:=evalf(Y +
h*subs({x=X,y=Y},g(x,y))); X:=X + h; pts[i]:=[X,Y];
od: y(x):=pts;
end;
> save EULER, "EULER.m":
```

We mention that the procedure EULER has been saved to a file with the extension .m. This means that the results are saved in *Maple internal format.* Maple internal format is only for the computer's use and one will not be able to see anything intelligible by viewing it or printing it out. This format is the same as that used for the files of Maple library. It can only be recalled by the Maple command read.

Note: To plot the points stored in an array, the array must be converted to a list. This may be accomplished with the Maple command seq: We illustrate this by the following example.

```
>   data:=EULER(x^2+y,[0,1.4],2/10,10);
```

$$data := pts$$

```
>   datalist:=seq(data[n],n=0..10);
```

$datalist := [0, 1.4], [\frac{1}{5}, 1.680000000], [\frac{2}{5}, 2.024000000], [\frac{3}{5}, 2.460800000],$

$[\frac{4}{5}, 3.024960000], [1, 3.757952000], [\frac{6}{5}, 4.709542400], [\frac{7}{5}, 5.939450880],$

$[\frac{8}{5}, 7.519341056], [\frac{9}{5}, 9.535209268], [2, 12.09025112]$

```
>   with(plots):
```

Warning, the name changecoords has been redefined

```
>   plot({datalist},style=POINT,symbol=BOX);
```

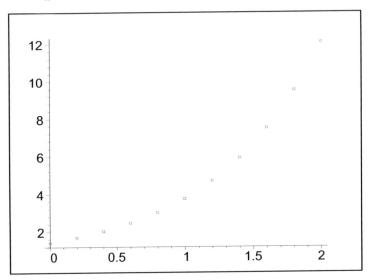

3.9   Computer Lab

In the study of fermentation kinetics, the *logistic equation*

$$\frac{dy}{dt} = k_1 y(1 - \frac{y}{k_2})$$

has been used frequently to describe the dynamics of cell growth, where k_1 and k_2 are positive constants. This equation is a modification of the *exponential growth equation*

$$\frac{dy}{dt} = k_1 y.$$

The term $(1 - y/k_2)$ in the logistic equation accounts for cessation of growth due to a limit nutrient.

- Find the exact solution of the logistic equation equation subject to the initial condition: $y(0) = y_0$, and determine the limiting value of the $y(t)$ as $t \to \infty$.

- Let the constants,

$$k_1 = 0.03120, \quad k_2 = 47.70, \quad y_0 = 5.0$$

be given, and assume that the time period is in the range

$0 \le t \le 10$. Apply the Euler, improved Euler and Runge-Kutta methods to the above IVP for $0 \le t \le 10$ by taking three different step sizes

$$h = 1, \quad h = 0.1, \quad h = 0.01.$$

Plot the exact solution curve and the numerical solution points in a figure with suitable symbols.

- Print the exact solution and the approximations at $t = 2, 4, 6, 8$, and 10.

- Compute the absolute errors:

$$E_h = |y(10) - y_h| \quad \text{for} \quad h = 1, \quad 0.1, \quad 0.01,$$

where y_h stands for the corresponding approximate solutions at $t = 10$.

- Plot E_h vs h on a log-log scale and employ the method of the least squares to find the orders of convergence for the three methods, respectively.

3.10 Supplementary Maple Programs

3.10.1 The Order of Convergence

The following is a procedure for computing the equation of the least squares power curve from the discrete data

```
>  with(linalg):
```

Warning, the protected names norm and trace have been redefined and unprotected

```
>  L:=proc(mesh,absE,N)
>  local x,y,a11,a12, b1, b2, i, sys, sol, slope, c, A, b ;
>  a11:=0; a12:=0; b1:=0; b2:=0;
>  for i to N do
>  x:=ln(mesh[i]); y:=ln(absE[i]);
>  a11:=a11+x^2; a12:=a12+x;
>  b1:=b1+x*y; b2:=b2+y;
>  od;
>  A:=<<a11,a12>|<a12,N>>;
>  b:=<b1,b2>;
```

```
>   sol:=linsolve(A,b);
>   slope:=sol[1]; c:=exp(sol[2]);
>   c*h^(slope); end;
```

$$L := \mathbf{proc}(mesh,\ absE,\ N)$$
$$\mathbf{local}\ x,\ y,\ a11,\ a12,\ b1,\ b2,\ i,\ sys,\ sol,\ slope,\ c,\ A,\ b;$$
$$a11 := 0;$$
$$a12 := 0;$$
$$b1 := 0;$$
$$b2 := 0;$$
$$\mathbf{for}\ i\ \mathbf{to}\ N\ \mathbf{do}$$
$$x := \ln(mesh_i);$$
$$y := \ln(absE_i);$$
$$a11 := a11 + x^2;$$
$$a12 := a12 + x;$$
$$b1 := b1 + x*y;$$
$$b2 := b2 + y$$
$$\mathbf{end\ do};$$
$$A := \langle\langle a11,\ a12\rangle\,|\,\langle a12,\ N\rangle\rangle;$$
$$b := \langle b1,\ b2\rangle;$$
$$sol := \mathrm{linsolve}(A,\ b);$$
$$slope := sol_1;$$
$$c := \exp(sol_2);$$
$$c * h^{slope}$$
$$\mathbf{end\ proc}$$

We now the following example to test the procedure.

$$dy/dt = 1 - t + 4y, \quad y(0) = 1.$$

Numerical Results are obtained from the Euler method.

```
>   deq:=diff(y(t),t)=1 -t + 4*y(t);
```

$$deq := \frac{\partial}{\partial t}\,y(t) = 1 - t + 4\,y(t)$$

```
>   ic:=y(0)=1;
```

$$ic := y(0) = 1$$

```
>   with(DEtools):
```

Warning, the name adjoint has been redefined

```
>   dsolve({deq,ic},y(t));
```

$$y(t) = -\frac{3}{16} + \frac{1}{4}\,t + \frac{19}{16}\,e^{(4\,t)}$$

```
>   read "EULER.m":
>   Y1:=EULER(1-x+4*y,[0,1],0.1,10);
```

$$Y1 := pts$$

```
>  Y1[10];
```

$$[1.0,\ 34.41149028]$$

```
>  Y2:=EULER(1-x+4*y,[0,1],0.05,20);
```

$$Y2 := pts$$

```
>  Y2[20];
```

$$[1.00,\ 45.58839990]$$

```
>  Y3:=EULER(1-x+4*y,[0,1],0.025,40);
```

$$Y3 := pts$$

```
>  Y3[40];
```

$$[1.000,\ 53.80786596]$$

```
>  Y4:=EULER(1-x+4*y,[0,1],0.01,100);
```

$$Y4 := pts$$

```
>  Y4[100];
```

$$[1.00,\ 60.03712596]$$

```
>  soln:=dsolve({deq,ic},y(t));
```

$$soln := y(t) = -\frac{3}{16} + \frac{1}{4}\,t + \frac{19}{16}\,e^{(4\,t)}$$

```
>  ex:=subs(t=1,rhs(soln));
```

$$ex := \frac{1}{16} + \frac{19}{16}\,e^4$$

```
>  ex:=evalf(%);
```

$$ex := 64.89780316$$

```
>  ex:=64.89780316; h:=array(1..4);
```

$$ex := 64.89780316$$
$$h := \operatorname{array}(1..4,\ [])$$

```
>  h[1]:=0.1; Error[1]:=abs(ex-op(2,Y1[10]));
```

$$h_1 := .1$$
$$Error_1 := 30.48631288$$

```
>  h[2]:=0.05; Error[2]:=abs(ex-op(2,Y2[20]));
```

$$h_2 := .05$$
$$Error_2 := 19.30940326$$

```
>  h[3]:=0.025; Error[3]:=abs(ex-op(2,Y3[40]));
```

$$h_3 := .025$$
$$Error_3 := 11.08993720$$

```
>  h[4]:=0.01; Error[4]:=abs(ex-op(2,Y4[100]));
```

$$h_4 := .01$$
$$Error_4 := 4.86067720$$

```
>  leqn:=L(h,Error,4);
```

$$leqn := 203.1817192\, h^{.8012818552}$$

```
>  with(plots):
```

Warning, the name changecoords has been redefined
```
>  plot1:=loglogplot([seq([h[i],Error[i]],i=1..4)],
>  style=POINT,symbol=CIRCLE):
>  plot2:=loglogplot(leqn,h=0.01..0.1,'-E-'=4.0..32.0):
>  display({plot1,plot2},title='Data
>  points vs Least Squares
>  Curve');
```

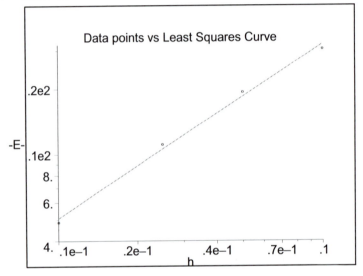

We remark that the slope of the least squares curve (i.e., $m = 0.8012818552$ in the present example) gives the order of convergence of the numerical scheme.

Chapter 4

Differential Equations with Constant Coefficients

4.1 Second-Order Equations with Constant Coefficients

We consider the second-order, linear, homogeneous, constant coefficient equation

$$a\frac{d^2y}{dt^2} + b\frac{dy}{dt} + cy = 0.$$

We enter this equation first into MAPLE and then use dsolve to find the form of the solution, namely

```
>   ode1:=a*diff(y(t),t$2)+b*diff(y(t),t)+c*y(t);
```

$$ode1 := a\left(\frac{\partial^2}{\partial t^2}y(t)\right) + b\left(\frac{\partial}{\partial t}y(t)\right) + cy(t)$$

```
>   dsolve(ode1,y(t));
```

$$y(t) = \_C1\,e^{(-1/2\,\frac{(b-\sqrt{b^2-4\,a\,c})\,t}{a})} + \_C2\,e^{(-1/2\,\frac{(b+\sqrt{b^2-4\,a\,c})\,t}{a})}$$

This is the solution; however, it takes different form depending on whether $b^2 - 4ac$ is positive, negative, or zero. When this expression, the discriminant, is positive we have two real solutions, when negative we get two complex solutions which are complex conjugates of each other, and when zero we find the second solution is t multiplied by the first solution. We can summarize this statement with the following MAPLE program that solves the homogeneous, constant coefficient, linear, second-order differential equation. We do this by means of nested if statements. The first condition determines whether $b^2 - 4ac > 0$. If this is so we get the general solution having two real but different roots to the characteristic equation. If this is not the case there are still two possibilities, namely $b^2 - 4ac = 0$, in which case we have the repeated real roots for the characteristic equation, or $b^2 - 4ac < 0$ which corresponds to complex conjugate roots. The function const_coeff(a,b,c,t) distinguishes which of these cases holds, and upon simplifying gives the general solution

```
>   with(plots):with(DEtools):
```

Warning, the name changecoords has been redefined

```
>   const_coef:=proc(a,b,c,t) description
>   "This program computes the solution of the homogeneous,
>   second-order, linear differential equation with constant
>   coefficients"; if b^2-4*a*c >0 then
>   _C1*exp((-b+sqrt(b^2-4*a*c))*t/(2*a))+
>   _C2*exp((-b-sqrt(b^2-4*a*c))*t/(2*a)) elif b^2-4*a*c =0 then
>   _C1*exp(-b*t/(2*a))+ _C2*t*exp(-b*t/(2*a))
>   else exp(-b*t/(2*a))*( _C1*sin(sqrt(4*a*c-b^2)*t/(2*a)) +
>   _C2*cos(sqrt(4*a*c-b^2)*t/(2*a)));
>   fi; end;
```

$const\_coef := \textbf{proc}(a, b, c, t)$

description "This program computes the solution of the homogeneous, second-order, linear differential equation with constant coefficients";

$\textbf{if}\, 0 < b^2 - 4 * c * a\, \textbf{then}\, \_C1 * \exp(1/2 * (-b + \text{sqrt}(b^2 - 4 * c * a)) * t/a)$

$+ \_C2 * \exp(1/2 * (-b - \text{sqrt}(b^2 - 4 * c * a)) * t/a)$

$\textbf{elif}\, b^2 - 4 * c * a = 0\, \textbf{then}\, \_C1 * \exp(-1/2 * b * t/a) +$

$\_C2 * t * \exp(-1/2 * b * t/a)$

$\textbf{else}\, \exp(-1/2 * b * t/a) *$

$(\_C1 * \sin(1/2 * \text{sqrt}(4 * c * a - b^2) * t/a) + \_C2 * \cos(1/2$

$* \text{sqrt}(4 * c * a - b^2) * t/a))$

end if

end proc

Let us test this function on several examples and then check these results with dsolve

```
>   const_coef(1,0,4,t);
```

$$\_C1\, \sin(2t) + \_C2\, \cos(2t)$$

Now we enter the differential equation and use dsolve

```
>   dsolve(diff(y(t),t$2)+4*y(t),y(t));
```

$$y(t) = \_C1\, \sin(2t) + \_C2\, \cos(2t)$$

which is the same result. We test the method with another example

```
>   const_coef(2,2,4,t);
```

$$e^{-t/2}\left(\_C1\, \sin\left(1/2\, \sqrt{7}t\right) + \_C2\, \cos\left(1/2\, \sqrt{7}t\right)\right)$$

Once again we see that dsolve gives the same answer

```
>   dsolve(2*diff(y(t),t$2)+2*diff(y(t),t)+4*y(t)
>   ,
>   y(t));
```

$$y(t) = \_C1\, e^{(-1/2\,t)} \sin(\frac{1}{2}\sqrt{7}t) + \_C2\, e^{(-1/2\,t)} \cos(\frac{1}{2}\sqrt{7}t)$$

Several other examples follow

```
>  const_coef(2,4,1,t);
```

$$\_C1\, e^{(1/2\,(-2+\sqrt{2})\,t)} + \_C2\, e^{(-1/2\,(2+\sqrt{2})\,t)}$$

```
>  const_coef(1,2,0,t);
```

$$\_C1 + \_C2\, e^{(-2\,t)}$$

```
>  dsolve(diff(y(t),t$2)+2*diff(y(t),t),
>  y(t));
```

$$y(t) = \_C1 + \_C2\, e^{(-2\,t)}$$

4.2   Variation of Parameters

4.2.1   The Wronskian

We recall that for second-order, ordinary differential equations the solutions are **linearly independent** on the interval $\mathcal{I}$ if the **Wronskian determinant**

$$\begin{vmatrix} \varphi_1 & \varphi_2 \\ \varphi_1' & \varphi_2' \end{vmatrix}$$

does <u>not vanish</u> there.

```
>  with(linalg):
```

Warning, the protected names norm and trace have been redefined
and unprotected

First we load DEtools and linalg before we begin

```
>  with(DEtools):
>  with(linalg):
```

We introduce our own procedure for the Wronskian matrix for a second-order, linear differential equation. We call the procedure W and it depends on the two solutions $y_1(x)$, $y_2(x)$ of the differential equation.

```
>  my_wronskian :=proc(y1,y2,x)
>  simplify(det(linalg[matrix](2,2,[y1,y2,diff(y1,x),diff(y2,x)])) )
>  end proc;
```

$$my\_wronskian := \mathbf{proc}(y1,\, y2,\, x)$$
$$\text{simplify}(\det(linalg_{matrix}(2,\, 2,\, [y1,\, y2,\, \text{diff}(y1,\, x),\, \text{diff}(y2,\, x)])))$$
$$\mathbf{end\ proc}$$

For the differential equation

$$\frac{d^2 y}{dx^2} + y = 0$$

we know there are two solutions $\sin(x)$ and $\cos(x)$. Let us check whether they are independent by evaluating the Wronskian determinant.

```
>  my_wronskian(sin(x),cos(x),x);
```
$$-1$$

We recall that we can solve constant coefficient, homogeneous equations using the **characteristic equation**. Let us consider a general differential equation with constant coefficients

$$a\frac{d^2y(t)}{dt^2} + b\frac{d\,y(t)}{dt} + cy(t) = 0$$

The characteristic equation is

```
>   a*m^2+b*m+c;
```
$$a\,m^2 + b\,m + c$$

```
>   solve(a*m^2+b*m+c,m)
>   ;
```
$$\frac{1}{2}\frac{-b+\sqrt{b^2-4\,a\,c}}{a},\ \frac{1}{2}\frac{-b-\sqrt{b^2-4\,a\,c}}{a}$$

Let us introduce the corresponding solutions in exponential form.

```
>   u1 :=exp(x/(2*a)*(-b+sqrt(b^2-4*a*c)));
```
$$u1 := e^{(\frac{x\,(-b+\sqrt{b^2-4\,a\,c})}{2a})}$$

```
>   u2 :=exp(x/(2*a)*(-b-sqrt(b^2-4*a*c)));
```
$$u2 := e^{(\frac{x\,(-b-\sqrt{b^2-4\,a\,c})}{2a})}$$

Without concerning ourselves with the natures of the roots of the characteristic equation let us formally compute the Wronskian.

```
>   my_wronskian(u1,u2,x);
```

$$\left[\left[e^{(1/2\frac{x\,(-b+\sqrt{b^2-4\,a\,c})}{a})},\ e^{(1/2\frac{x\,(-b-\sqrt{b^2-4\,a\,c})}{a})} \right], \right.$$
$$\left[\frac{1}{2}\frac{(-b+\sqrt{b^2-4\,a\,c})\,e^{(1/2\frac{x\,(-b+\sqrt{b^2-4\,a\,c})}{a})}}{a}, \right.$$
$$\left. \left. \frac{1}{2}\frac{(-b-\sqrt{b^2-4\,a\,c})\,e^{(1/2\frac{x\,(-b-\sqrt{b^2-4\,a\,c})}{a})}}{a} \right] \right]$$

Another way to calculate the Wronskian is to use the MAPLE command Wronskian and the Vector Calculus package. We do that in the following:

```
>   with(VectorCalculus);
```

Define:
$$\alpha = e^{(1/2)*x*(-b+\sqrt{-4*a*c+b^2})/a}$$
$$\beta = e^{(1/2)*x*(-b-\sqrt{-4*a*c+b^2})/a}$$

$$\gamma = (1/2)*(-b+\sqrt{-4*a*c+b^2})/a*e^{(1/2)*x*(-b+\sqrt{-4*a*c+b^2})/a}$$
$$\delta = (1/2)*(-b-\sqrt{-4*a*c+b^2})/a*e^{(1/2)*x*(-b-\sqrt{-4*a*c+b^2})/a}$$
then,

```
>  Wronskian([u1, u2], x, 'determinant');
```

$$\begin{bmatrix} \alpha & \beta \\ \gamma & \delta \end{bmatrix}, -e^{x*(-b+\sqrt{-4*a*c+b^2})/(2*a)}*e^{-x*(b+\sqrt{-4*a*c+b^2})/(2*a)}$$
$$*\sqrt{-4*a*c+b^2}/a$$

$$-\frac{\sqrt{-4\,ac+b^2}}{a}e^{\frac{x\left(-b+\sqrt{-4\,ac+b^2}\right)}{2\,a}}e^{-\frac{x\left(b+\sqrt{-4\,ac+b^2}\right)}{2\,a}}$$

We can see that providing $b^2 - 4ac \neq 0$ the Wronskian determinant does not vanish. If this term does vanish we have a double root to the characteristic equation and we do not obtain two **distinct** $\{e^{mx}, e^{mx}\}$, and hence, independent solutions. However, in the case where there are distinct roots of the characteristic equation the Wronskian cannot vanish and there are two independent solutions.

Let us compute the Wronskian for the differential equation,[1]

$$\frac{d^2 y(t)}{dt^2} + 2\frac{d\,y(t)}{dt} + 2y(t) = 0.$$

We are required to find two linearly independent solutions of the equation; hence we consider the characteristic equation

```
>  m^2+2*m+2;
```

$$m^2 + 2m + 2$$

and solve it to get the roots.

```
>  solve(m^2+2*m+2,m);
```

$$-1 + I, -1 - I$$

```
>  i:=sqrt(-1);
```

$$i := I$$

Let us compute the Wronskian of these two solutions.

```
>  my_wronskian(exp(x*(-1+I)),exp(x*(-1-I)),x);
```

$$-2\,I\,e^{(-2\,x)}$$

Since the Wronskian can only vanish at $x = \infty$ we conclude that the general solution to this equation is given by

```
>  y(x):= c_1*exp(x*(-1+i)) +c_2*exp(x*(-1-i))
>  ;
```

$$1 = c\_1 + c\_2, \, 0 = (-1+i)\,c\_1 - (1+i)\,c\_2$$

---

[1] Recall that by Abel's theorem any Wronskian is found up to a multiplicative constant by finding two independent solutions.

Suppose we wish to find the solution satisfying the initial conditions $y(0) = 1$, $y'(0) = 0$. Since the Wronskian is nowhere zero we know this is possible. We are then led to solve the equations for the coefficients c_1 and c_2.

```
>   1= c_1 +c_2,  0= c_1*(-1+i) +c_2*(-1-i) ;
```

$$1 = c\_1 + c\_2, 0 = (-1 + I)\,c\_1 + (-1 - I)\,c\_2$$

```
>   solve({1= c_1 +c_2,  0= c_1*(-1+i)
>   +c_2*(-1-i)},{c_1,c_2});
```

$$\{c\_2 = \frac{1}{2} + \frac{1}{2}\,I,\ c\_1 = \frac{1}{2} - \frac{1}{2}\,I\}$$

$$\{c\_1 = 1/2 - i/2, c\_2 = 1/2 + i/2\}$$

$$(\frac{1}{2} - \frac{1}{2}\,I)\,e^{x(-1+I)} + (\frac{1}{2} + \frac{1}{2}\,I)\,e^{x(-1-I)}$$

We can write a program that solves the initial value problem for a homogeneous, second-order, linear ordinary differential equation. The first program IVP depends on four parameters, the initial point $x = x_0$, the two initial values values $y(x_0) = a$, $\frac{dy}{dx}(x_0) = b$, and two known independent solutions. The program computes the coefficients that appear in the general solution.

```
>   IVP :=proc(x0,a,b,y1,y2) local eq1,
>   eq2,C1,C2,A; eq1:=simplify(subs(x=x0,C1*y1+C2*y2=a));
>   eq2:=simplify(subs(x=x0,C1*diff(y1,x)+C2*diff(y2,x)=b));
>   solve({eq1,eq2},{C1,C2});
>   end proc;
```

$IVP := \mathbf{proc}(x0,\ a,\ b,\ y1,\ y2)$
$\mathbf{local}\ eq1,\ eq2,\ C1,\ C2,\ A;$
  $eq1 := \mathrm{simplify}(\mathrm{subs}(x = x0,\ C1 * y1 + C2 * y2 = a));$
  $eq2 := \mathrm{simplify}(\mathrm{subs}(x = x0,\ C1 * \mathrm{diff}(y1,\ x) + C2 * \mathrm{diff}(y2,\ x) = b));$
  $\mathrm{solve}(\{eq1,\ eq2\},\ \{C1,\ C2\})$
end proc

We test this program on the differential equation we have just been discussing

$$\frac{d^2 y(t)}{dt^2} + 2\frac{d\,y(t)}{dt} + 2y(t) = 0.$$

We have already computed two independent solutions, namely $\exp x(-i + 1)$ and $\exp x(-i - 1)$.

```
>   IVP(0,1,0,exp(x*(-1+I)),exp(x*(-1-I)));
```

$$\{C1 = 1/2 - i/2, C2 = 1/2 + i/2\}$$

```
>   IVP_soln :=proc(x0,a,b,y1,y2) local eq1,
>   eq2,C1,C2,A,B; eq1:=simplify(subs(x=x0,C1*y1+C2*y2=a));
>   eq2:=simplify(subs(x=x0,C1*diff(y1,x)+C2*diff(y2,x)=b));
>   A:=C_1*y1+C_2*y2; B:=solve({eq1,eq2},{C1,C2});
>   print({op(2,B),op(1,B)});
>   subs({C_1=op(2,B),C_2=op(1,B)},A);end proc;
```

   IVP_soln := **proc**(*x0, a, b, y1, y2*)
   local *eq1, eq2*, *C1, C2, A, B*;
      eq1 := simplify(subs(*x* = *x0*, *C1* ∗ *y1* + *C2* ∗ *y2* = *a*));
      eq2 := simplify(subs(*x* = *x0*, *C1* ∗ diff(*y1, x*) + *C2* ∗ diff(*y2, x*) = *b*));
      A := *C_1* ∗ *y1* + *C_2* ∗ *y2* ;
      B := solve({*eq1, eq2*}, {*C1, C2*});
      print({op(1, *B*), op(2, *B*)});
      subs({*C_2* = op(1, *B*), *C_1* = op(2, *B*)}, *A*)
   end proc

The next program substitutes the found coefficients for those in the representation of the general solution.

```
>   IVP_soln(0,1,0,exp(x*(-1+I)),exp(x*(-1-I)));
```

$$\{C2 = \frac{1}{2} + \frac{1}{2} I, \ C1 = \frac{1}{2} - \frac{1}{2} I\}$$

$$C1 \, e^{((-1+I)x)} + C2 \, e^{((-1-I)x)} = (\frac{1}{2} - \frac{1}{2} I) \, e^{((-1+I)x)} + (\frac{1}{2} + \frac{1}{2} I) \, e^{((-1-I)x)}$$

which is exactly what we just computed in desk calculator mode DCM .

```
>   simplify(%);
```

$$e^{(-1+i)x} \, C2 + e^{(-1-i)x} \, C1 = (1/2 + i/2) \, e^{(-1+i)x} + (1/2 - i/2) \, e^{(-1-i)x}$$

MAPLE can solve directly initial value problems. With this in mind we input the differential equation and the initial conditions.

```
>   ode1:=diff(y(x),x$2)+2*diff(y(x),x)+2*y(x)=0;inits:=y(0)=1,D(y)(0)=0;
```

$$ode1 := (\frac{\partial^2}{\partial x^2} \, y(x)) + 2 \, (\frac{\partial}{\partial x} \, y(x)) + 2 \, y(x) = 0$$

$$inits := y(0) = 1, \ D(y)(0) = 0$$

We then solve the initial value problem using `dsolve`.

```
>   dsolve({ode1,inits},y(x));
```

$$y(x) = e^{(-x)} \sin(x) + e^{(-x)} \cos(x)$$

which appears different from the solution we obtained before. Let us check this by subtracting one solution from the other and `simplify`.

```
>   exp(-x)*sin(x)+exp(-x)*cos(x)-((1/2-1/2*I)*exp((-1+I)*x)
>   +(1/2+1/2*I)*
>   exp((-1-I)*x));
```

$$e^{(-x)} \sin(x) + e^{(-x)} \cos(x) - (\frac{1}{2} - \frac{1}{2} I) \, e^{((-1+I)x)} - (\frac{1}{2} + \frac{1}{2} I) \, e^{((-1-I)x)}$$

To check whether these two are the same we `convert` to `trigonometric` terms and `simplify`. We see that they are indeed the same.

```
>   simplify(convert(%,trig));
```

$$0$$

```
>   with(linalg):with(DEtools):
```

```
Warning, the protected names norm and trace have been redefined
and unprotected
```

```
Warning, the name adjoint has been redefined
```

We now write a program based on the variation of parameters formula. It is required that we have two solutions of the homogeneous equation. We have already given a program to do this in the constant coefficient case.

```
>  Vari_Param2 :=proc(f,y1,y2) local W,W1,
>  W2: description "This program uses the method of variation of
>  parameters to compute a particular solution of a second order,
>  nonhomogeneous equation.  Knowledge of two independent solutions
>  of the homogeneous equation are required";
>  W:=simplify(det(linalg[matrix](2,2,
>  [y1,y2,diff(y1,x),diff(y2,x)]))); W1:=y2*f; W2:=y1*f;
>  simplify(-y1*int(W1/W,x)
>  +y2*int(W2/W,x)) end proc;
```

$Vari\_Param2 := \mathbf{proc}(f, y1, y2)$

$\mathbf{local}\, W, W1, W2;$

$\mathbf{description}$ "This program uses the method of variation of parameters to computer a particular solution of a second order,\ nnonhomogeneous equation. Knowledge of two independent solutions of the homogeneous equation are required";

$W := \text{simplify}(\det(linalg_{matrix}(2, 2, [y1, y2, \text{diff}(y1, x), \text{diff}(y2, x)])));$

$W1 := y2 * f\,;$

$W2 := f * y1\,;$

$\text{simplify}(-y1 * \text{int}(W1/W, x) + y2 * \text{int}(W2/W, x))$

$\mathbf{end\ proc}$

Let us test our program on the differential equation

$$\frac{d^2\, y(x)}{dx^2} + 2\frac{d\, y(x)}{dx} + y(x) = cos(x).$$

```
>  Vari_Param2(cos(x),exp(-x),x*exp(-x));
```

$$1/2 \sin(x)$$

```
>  ode1:=diff(y(x),x$2)+2*diff(y(x),x)+y(x)=cos(x);
```

$$ode1 := (\tfrac{\partial^2}{\partial x^2}\, y(x)) + 2\,(\tfrac{\partial}{\partial x}\, y(x)) + y(x) = cos(x)$$

We now check to see if this is correct by using dsolve.

```
>  dsolve(ode1,y(x));
```

$$y(x) = e^{-x}\_C2 + xe^{-x}\_C1 + 1/2 \sin(x)$$

The term without an arbitrary constant must be a particular solution and it is exactly the term we found. We test the program on another differential equation.

```
> ode2:=diff(y(x),x$2)+2*diff(y(x),x)+y(x)=x^2*exp(-x);
```

$$ode2 := (\tfrac{\partial^2}{\partial x^2}\, \mathrm{y}(x)) + 2\,(\tfrac{\partial}{\partial x}\, \mathrm{y}(x)) + \mathrm{y}(x) = x^2\, e^{(-x)}$$

```
> Vari_Param2(x^2*exp(-x),exp(-x),x*exp(-x));
```

$$\frac{1}{12}\, e^{(-x)}\, x^4$$

Once more we get the correct result.

```
> dsolve(ode2,y(x));
```

$$\mathrm{y}(x) = e^{(-x)}\, \_C2 + x\, e^{(-x)}\, \_C1 + \frac{1}{12}\, e^{(-x)}\, x^4$$

```
> const_coef(1,1,1,t);
```

$$e^{(-1/2\,t)}\,(\_C1\,\sin(\tfrac{1}{2}\,\sqrt{3}\,t) + \_C2\,\cos(\tfrac{1}{2}\,\sqrt{3}\,t))$$

```
> const_coef(1,0,1,t);
```

$$\_C1\,\sin(t) + \_C2\,\cos(t)$$

Project Write a program that finds the general solution to a second-order, linear, nonhomogeneous, ordinary differential equation with constant coefficients. Incorporate the programs `const_coef` and `Var_Param_2` to do this.

4.3   The Method of Undetermined Coefficients

In this section, we consider the case of differential equations whose left-hand-side has constant coeffiecients and non-homogeneous term is of a special form. More precisely we consider the equations

$$ay'' + by' + cy = F(t),$$

where the nonhomogeneous term F has the form

$$F(t) = \begin{cases} p_n(t) \\ p_n(t)e^{at} \\ p_n(t)\sin t \\ p_n(t)\cos t \end{cases},$$

where $p_n(t)$ is a polynomial of degree n. A particular solution of the differential equation (4.3) with right-hand-side (4.3) can be found in the forms

$$t^m P_n(t)$$
$$t^m P_n(t)e^{at}$$
$$t^s\,(P_n(t)\sin t + t^s Q_n(t)\cos t)$$
$$t^s\,(P_n(t)\sin t + t^s Q_n(t)\cos t)$$

respectively. Here $P_n(t)$ and $Q_n(t)$ are polynomials of degree n which are to be determined, and s is an integer to be determined. The integer s is chosen so that <u>no term in the trial solution</u> is a solution of the homogeneous equation. We consider the homogeneous equation

$$a\,\frac{d^2y}{dt^2} + b,\frac{dy}{dt} + cy = f(x),$$

where a, b, and c are constants and the non-homogeneous term $f(x)$ is continuous.

```
>  with(DEtools):
```

In this session we write a procedure which analyses the solution of the equation(4.3). The procedure will depend on the coefficents a, b and c. Moreover, it will depend on the initial data

$$y(0) = y_0, \text{ and } \frac{d\,y(0)}{dt} = y_1.$$

We recall the program `const_coeff(a,b,c,t)` and use this to investigate bon-homogeneous, second-order, linear equations. As an example, we consider

$$\frac{d^2\,y(t)}{dt^2} + 2\frac{d\,y(t)}{dt} + \vec{y}(t) = cos(t),$$

that is we set $a = 1$, $b = 2$, $c = 1$, and $f(t) := cos(t)$, into`const_coeff(a,b,c,t)`. First we input the equation as ode1.

```
>  ode1:=diff(y(t),t$2)+2*diff(y(t),t)+y(t)=cos(t);
```

$$ode1 := (\tfrac{\partial^2}{\partial t^2}\,y(t)) + 2\,(\tfrac{\partial}{\partial t}\,y(t)) + y(t) = cos(t)$$

Our program computes the general solution to the homogeneous problem as

```
>  const_coef(1,2,1,t);
```

$$\_C1\,e^{(-t)} + \_C2\,t\,e^{(-t)}$$

Since the non-homogeneous term is not included in the solution set of the homogeneous equation we choose a particular solution in the form $y(t) = A\cos(t) + B\sin(t)$ and substitute this into the differential equation.

```
>  subs(y(t)=A*cos(t)+B*sin(t),ode1);
```

$$(\tfrac{\partial^2}{\partial t^2}\,(A\cos(t) + B\sin(t))) + 2\,(\tfrac{\partial}{\partial t}\,(A\cos(t) + B\sin(t))) + A\cos(t) + B\sin(t) = cos(t)$$

```
>  eq1:=simplify(%);
```

$$eq1 := -2\,A\sin(t) + 2\,B\cos(t) = cos(t)$$

Next we determine the coefficients A and B so that this choice makes $y(x)$ a solution.

```
>  lhs(%);
```

$$-2\,A\sin(t) + 2\,B\cos(t)$$

Hence, we obtain a particular solution of the form

$$y\_p := t \mapsto 1/2 \sin(t)$$

and the general solution is then seen to be

$$y\_g := t \mapsto \_C1\, e^{-t} + \_C2\, t e^{-t} + 1/2 \sin(t)$$

We now seek the solution which has the initial conditions

$$y(0) = \frac{dy}{dx}(0) = 0,$$

which we input as follows for later use.

> inits:=y(0)=0,D(y)(0)=0;

$$inits := y(0) = 0,\ D(y)(0) = 0$$

We now compute the solution and its derivative and evaluate these to 0 at $x = 0$.

> simplify(subs(t=0,diff(y_g(t),t)));

$$-\_C1 + \_C2 + \frac{1}{2}$$

> simplify(subs(t=0,y_g(t)));

$$\_C1$$

Next we fix the arbitrary coefficients so that the initial conditions are satisfied.

> solve({_C1,-_C1+_C2+1/2},{_C1,_C2});

$$\{\_C1 = 0,\ \_C2 = \frac{-1}{2}\}$$

> u(t):=subs({_C1 = 0, _C2 = -1/2},y_g(t));

$$u(t) := -\frac{1}{2} t\, e^{(-t)} + \frac{1}{2} \sin(t)$$

We shall next use MAPLE to compute the solution to the initial value problem directly. We first ask a suggestion from MAPLE, and receive the following.

> odeadvisor(ode1);

$$[[\_2nd\_order,\ \_linear,\ \_nonhomogeneous]]$$

> dsolve({ode1,inits},y(t));

$$y(t) = -\frac{1}{2} t\, e^{(-t)} + \frac{1}{2} \sin(t)$$

Hence, MAPLE gave us the same answer as we had constructed. As we want to use the same notation we unassume the previous unknown.

> unassume(y(t));

$$unassume(y(t))$$

In the next example we consider a nonhomogeneous term that is contained in the solution space of the non-homogeneous equations solutions.

> ode2:=diff(y(x),x$2)+2*diff(y(x),x)+y(x)=x^2*exp(-x);

$$ode2 := (\tfrac{\partial^2}{\partial x^2}\, y(x)) + 2\,(\tfrac{\partial}{\partial x}\, y(x)) + y(x) = x^2\, e^{(-x)}$$

If the right-hand side were not in the solution space we would choose the trial solution in the form

$$y(x) \;=\; Ax^2 \,+\, Bx \,+\, C)\exp{-x};$$

however, to remove each term in this expression from the solution space we must multiply this term by x^2.

```
>   subs(y(x)=x^2*(A*x^2+B*x+C)*exp(-x),ode2);
```

$$(\tfrac{\partial^2}{\partial x^2}\, x^2\,(A\,x^2 + B\,x + C)\, e^{(-x)}) + 2\,(\tfrac{\partial}{\partial x}\, x^2\,(A\,x^2 + B\,x + C)\, e^{(-x)})$$
$$+ x^2\,(A\,x^2 + B\,x + C)\, e^{(-x)} = x^2\, e^{(-x)}$$

```
>   E:=simplify(%);
```

$$E \;:=\; \left(12\,Ax^2 + 6\,Bx + 2\,C\right)e^{-x} = x^2 e^{-x}$$

We now seek to determine the coefficients A, B and C.

```
>   F:=simplify(E*exp(x));
```

$$F := 12\,A\,x^2 + 6\,B\,x + 2\,C = x^2$$

```
>   for k from 0 to 2 do
>   eq(k):=coeff(12*A*x^2+6*B*x+2*C+16*x^3*A+12*x^2*B+8*x*
>   C+4*x^4*A+4*x^3*
>   B+4*x^2*C-x^2,x,k) od;
```

$$eq(0) := 2\,C$$

$$eq(1) := 6\,B + 8\,C$$

$$eq(2) := 12\,A + 12\,B + 4\,C - 1$$

```
>   solve({eq(0),eq(1),eq(2)},{A,B,C});
```

$$\{C = 0,\; B = 0,\; A = \frac{1}{12}\}$$

The general solution now has the form

$$y\_gen := x \mapsto 1/12\,x^4 e^{-x} + \_C3\, e^{-x} + x\,\_C4\, e^{-x}$$

Finally we want the initial conditions to be satisfied

```
>   diff(y_gen(x),x);
```

$$\frac{1}{3}\,x^3\, e^{(-x)} - \frac{1}{12}\,x^4\, e^{(-x)} - \_C3\, e^{(-x)} + \_C4\, e^{(-x)} - x\,\_C4\, e^{(-x)}$$

```
>   eq3:=simplify(subs(x=0,y_gen(x)));
```

$$eq3 := \_C3$$

```
>   eq4:=simplify(subs(x=0,diff(y_gen(x),x)));
```

$$eq4 := -\,\_C3 + \_C4$$

```
>   solve({_C3,-_C3+_C4},{_C3,_C4});
```

$$\{\_C3 = 0,\; \_C4 = 0\}$$

Let us check this answer with dsolve.

```
>   dsolve({ode2,inits},y(x));
```

$$y(x) = \frac{1}{12} x^4 e^{(-x)}$$

The term $D := \sqrt{b^2 - 4ac}/2a$ determines the nature of the roots of the characteristic equation.

$$R := (a, b, c) \mapsto 1/2 \, \frac{\sqrt{-4\,ac + b^2}}{a}$$

```
>   D(1,2,1);
```

$$0$$

Hence, the example we just solved corresponds to a double root and there are two solutions $\exp -bt/a$, and $t \exp -bt/a$. Let us look for a more interesting case.

```
>   D(1,2,4);
```

$$I\sqrt{3}$$

```
>   with(DEtools):
>   ode3:=diff(y(t),t$2)+2*diff(y(t),t)+4*y(t)=cos(2*sqrt(3)*t);
```

$$ode3 := \left(\tfrac{\partial^2}{\partial t^2}\, y(t)\right) + 2\left(\tfrac{\partial}{\partial t}\, y(t)\right) + 4\,y(t) = \cos(2\sqrt{3}\,t)$$

```
>   dsolve({ode3,inits},y(t));
```

$$y(t) = -1/21\,e^{-t}\sqrt{3}\sin\left(\sqrt{3}t\right) + 1/14\,e^{-t}\cos\left(\sqrt{3}t\right) + 1/28\,\sqrt{3}\sin\left(2\sqrt{3}t\right)$$
$$-1/14\,\cos\left(2\sqrt{3}t\right)$$

```
>   ode4:=diff(y(t),t$2)+4*y(t)=cos(2*t);
```

$$ode4 := \left(\tfrac{\partial^2}{\partial t^2}\, y(t)\right) + 4\,y(t) = \cos(2\,t)$$

```
>   dsolve({ode4,inits},y(t));
```

$$y(t) = \frac{1}{4}\sin(2\,t)\,t$$

The above solution oscillates with larger and larger amplitudes, a feature of resonance. We plot this solution to illustrate the growth in amplitude.

```
>   with(plots):
```

Warning, the name changecoords has been redefined

```
>   plot(1/4*sin(2*t)*t,t=0..20*Pi);
```

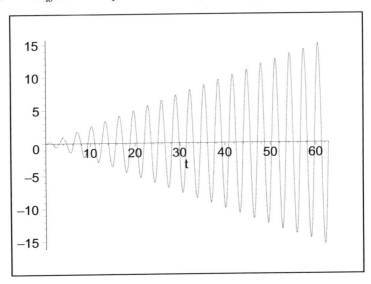

We solve and plot the differential equation with several variations on the nonhomogeneous term.

> with(DEtools): with(plots):

Warning, the name changecoords has been redefined

> inits:=y(0)=0,D(y)(0)=0;

$$inits := y(0) = 0, \text{D}(y)(0) = 0$$

> ode4:=diff(y(t),t$2)+4*y(t)=cos(2.1*t);

$$ode4 := \left(\frac{\partial^2}{\partial t^2}\, y(t)\right) + 4\, y(t) = \cos(2.1\, t)$$

> dsolve({ode4,inits},y(t));

$$y(t) = \frac{100}{41}\cos(2\, t) - \frac{100}{41}\cos\left(\frac{21}{10}\, t\right)$$

> plot(rhs(%),t=0..20*Pi);

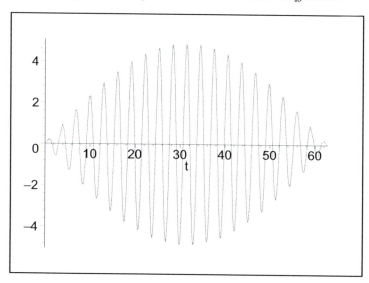

```
>   ode5:=diff(y(t),t$2)+4*y(t)=cos(2.4*t);
```

$$ode5 := (\tfrac{\partial^2}{\partial t^2}\, y(t)) + 4\, y(t) = \cos(2.4\, t)$$

```
>   dsolve({ode5,inits},y(t));
```

$$y(t) = \frac{25}{44}\cos(2\, t) - \frac{25}{44}\cos(\frac{12}{5}\, t)$$

```
>   plot(rhs(%),t=0..20*Pi);
```

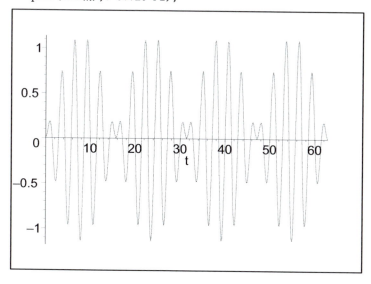

4.4   Higher-Order, Homogeneous Equations

In this session we solve higher-order homogeneous equations.

```
>   deq1:=12*diff(y(t),t$4)+31*diff(y(t),t$3)+75*
>   diff(y(t),t$2)+
>   37*diff(y(t),t)+5*y(t);
```

$$deq1 := 12\left(\frac{\partial^4}{\partial t^4}\,y(t)\right) + 31\left(\frac{\partial^3}{\partial t^3}\,y(t)\right) + 75\left(\frac{\partial^2}{\partial t^2}\,y(t)\right) + 37\left(\frac{\partial}{\partial t}\,y(t)\right) + 5\,y(t)$$

Here we input the characteristic equation.

```
>   solve(12*r^4+31*r^3+75*r^2+37*r+5);
```

$$\frac{-1}{3},\frac{-1}{4},\,-1-2\,I,\,-1+2\,I$$

Hence a general solution is of the form

$$C_1 \exp(-t/4) \;+\; C_2 \exp(-t/3) \;+\; C_3 \exp(-t)\cos(2t) \;+\; C_4 \exp(-t)\sin(2t).$$

```
>   with(DEtools):
>   dsolve(deq1,y(t));
```

$$y(t) = \_C1\,e^{(-1/4\,t)} + \_C2\,e^{(-1/3\,t)} + \_C3\,e^{(-t)}\cos(2\,t) + \_C4\,e^{(-t)}\sin(2\,t)$$

We try a more complicated higher-order equation

```
>   deq2:=diff(y(t),t$4)+6*diff(y(t),t$3)+ ;
>   17*diff(y(t),t$2)+ 22*diff(y(t),t)+14*y(t);
```

$$deq2 := \left(\frac{\partial^4}{\partial t^4}\,y(t)\right) + 6\left(\frac{\partial^3}{\partial t^3}\,y(t)\right) + 17\left(\frac{\partial^2}{\partial t^2}\,y(t)\right) + 22\left(\frac{\partial}{\partial t}\,y(t)\right) + 14\,y(t)$$

We input the characteristic equation.

```
>   solve(r^4+6*r^3+17*r^2+22*r+14);
```

$$-1-I,\,-1+I,\,-2-I\sqrt{3},\,-2+I\sqrt{3}$$

The general solution must therefore have the form

$$C_1 \exp(-t)\sin(t) + C_2 \exp(-t)\cos(t) + C_3 \exp(-2t)\sin(\sqrt{3}\,t) + C_4 \exp(-2t)\cos(\sqrt{3}\,t).$$
$$(4.1)$$

We check this with Maple using dsolve.

```
> dsolve(deq2,y(t));
```

$$y(t) = \_C1\,e^{-2t}\sin\left(\sqrt{3}t\right) + \_C2\,e^{-2t}\cos\left(\sqrt{3}t\right) + \_C3\,e^{-t}\sin\left(t\right) + \_C4\,e^{-t}\cos\left(t\right)$$

```
> readlib(polar):
```

To find the roots of some characteristic equations it can be helpful to be able to use the de Moivre formula. For example if the characteristic equation requires finding a root of a real or complex number, such as, $r^6 + 1 = 0$, for example. Let us see how MAPLE solves some characteristic equations of this type. We consider first $r^2 = 1 - i$.

```
> solve(z^2=1-I);
```

$$\sqrt{1 - I},\ -\sqrt{1 - I}$$

The roots $\sqrt{1 - I},\ -\sqrt{1 - I}$ are really not satisfactor for us to work with. Instead, let us rewrite these in polar coordinates using `polar`.

```
> polar((1-I)^(1/2));
```

$$\text{polar}(2^{1/4},\ -\arctan(\frac{\sqrt{-2 + 2\sqrt{2}}}{\sqrt{2 + 2\sqrt{2}}}))$$

```
> polar(-(1-I)^(1/2));
```

$$\text{polar}(2^{1/4},\ -\arctan(\frac{\sqrt{-2 + 2\sqrt{2}}}{\sqrt{2 + 2\sqrt{2}}}) + \pi)$$

We next compute the four roots of -1.

```
> solve(r^4=-1);
```

$$1/2\,\sqrt{2} + i/2\sqrt{2},\ -1/2\,\sqrt{2} + i/2\sqrt{2},\ -1/2\,\sqrt{2} - i/2\sqrt{2},\ 1/2\,\sqrt{2} - i/2\sqrt{2}$$

```
> evalc((1-I)^(1/2));
```

$$\frac{1}{2}\,\sqrt{2 + 2\sqrt{2}} - \frac{1}{2}I\sqrt{-2 + 2\sqrt{2}}$$

```
> evalc(-(1-I)^(1/2));
```

$$-\frac{1}{2}\,\sqrt{2 + 2\sqrt{2}} + \frac{1}{2}I\sqrt{-2 + 2\sqrt{2}}$$

4.4.1 Polynomial Solutions

MAPLE can look for rational solutions, a fraction of two polynomials. If there are such solutions then one can construct the fundamental pair, or in the case of higher-order equations the fundamental n-tuplets more easily, if not directly. To search for rational solutions we first load `DEtools`.

```
> with(DEtools):
```

Let us seek rational solutions of the differential equation

```
> ode_4 := (D@@2)(z)(t) - 3/t*D(z)(t) +
> 3/t^2*z(t):
```

This is done with the use of `ratsols`

```
> ratsols(ode_4, z(t));
```

$$[t, t^3]$$

We test `ratsols` on another equation.

```
> ode_5:=diff(z(t),t$2)+2*diff(z(t),t)/t;
```

$$ode\_5 := (\tfrac{\partial^2}{\partial t^2}\, z(t)) + 2\,\frac{\tfrac{\partial}{\partial t}\, z(t)}{t}$$

```
> ratsols(ode_5, z(t));
```

$$[t^{-1}, 1]$$

This last equation only had a constant as a rational solution.

4.5   Nonhomogeneous Linear Equations

4.5.1   Undetermined Coefficients

We recall that the general solution to a nonhomogeneous, linear, differential equation can be written as a **particular** solution to the nonhomogeneous equation plus the general solution to the homogeneous equation. The Method of Undetermined Coefficients provides a means to construct a particular solution to nonhomogeous , linear, ordinary differential equations with constant coefficients[2] In this session we solve non-homogeneous, higher-order equations, using the variation of parameters method. We check our results using the MAPLE command `dsolve`.

```
> de1:=diff(y(t),t$3) + 4*diff(y(t),t)-t;
```

First let us note that the solution to the initial value problem can be obtained using `dsolve`. Note the syntax carefully!

$$de1 := (\frac{\partial^3}{\partial t^3}\, y(t)) + 4\,(\frac{\partial}{\partial t}\, y(t)) - t$$

```
> dsolve({de1,
> y(0)=0,D(y)(0)=0,D(D(y))(0)=1},y(t) );
```

$$y(t) = 3/16 - 3/16 \cos(2t) + 1/8\, t^2$$

We want to check MAPLEś result using the method of undetermined coefficients. To this end, we first solve the characteristic equation.

```
> solve(r^3+4*r);
```

---

[2]Actually the LHS may be of order n, that is we may consider differential equations
$$\sum_{k=0}^{n} a_k y^{(n-k)} = F(t)$$

$$0,\, 2\,I,\, -2\,I$$

Hence the general solution to the homogeneous equation is

$$C_1 + C_2 \cos(2t) + C_3 \sin(2t)$$

Now let us find a particular solution to the non-homogeneous equation. Since the non-homogeneous term is t the method of undetermined coefficients would usually suggest we use $at + b$; however, as 1 is a solution of the homogeneous equation; hence, we must use instead $y_p := at^2; + t$.

```
>  simplify(subs(y(t)=a*t^2+b*t,de1));
```

$$\frac{\partial^3}{\partial t^3}\left(at^2 + bt\right) + 4\,\frac{\partial}{\partial t}\left(at^2 + bt\right) - t$$

We solve for the two *undetermined* coefficients a and b

```
>  mapleinlineactive2dsimplify( ); 1
```

$$(8\,a - 1)\,t + 4\,b$$
$$\{a = 1/8, b = 0\}$$

This is exactly the non-homogeneous solution found by MAPLE.

$$z := t \mapsto C\_1 + C\_2 \cos(2\,t) + C\_3 \sin(2\,t) + 1/8\,t^2$$

We now want to calculate the value of the coefficients C_k, $k = 1, 2, 3$ in order to satisfy the initial conditions $y(0) = 0,\, y'(0) = 0,\, y''(0) = 1$.

```
>  eq1:=simplify(subs(t=0,C_1+C_2*cos(2*t)+C_3*s
>  in(2*t)+1/8*t^2));
```

$$eq1 := C\_1 + C\_2$$

```
>  eq2:=simplify(subs(t=0,diff(C_1+C_2*cos(2*t)+ ));
>  C_3*sin(2*t)+1/8*t^2,t) ));
```

$$eq2 := 2\,C\_3$$

```
>  eq3:=simplify(subs(t=0,diff(C_1+C_2*cos(2*t)+
>  C_3*sin(2*t)+1/8*t^2,t$2) )-1);
```

$$eq3 := -4\,C\_2 - \frac{3}{4}$$

```
>  solve({eq1,eq2,eq3},{C_1,C_2,C_3});
```

$$\{C\_1 = 3/16,\, C\_2 = -3/16,\, C\_3 = 0\}$$

So our result agrees with that of MAPLE We consider the above differential equation but with a different non-homogeneous term, namely $\sin(2t)$. This term is also a solution of the homogeneous equation so instead of using $a\sin(2t) + b\cos(2t)$, we need to use $t\,(a\sin(2t) + b\cos(2t))$,

```
> de2:=diff(y(t),t$3) +
> 4*diff(y(t),t)-sin(2*t);
```

$$de2 := (\frac{\partial^3}{\partial t^3} y(t)) + 4 (\frac{\partial}{\partial t} y(t)) - \sin(2t)$$

We want to

```
> simplify(subs(y(t)=t*(A*cos(2*t)+B*sin(2*t)),
> de2));
```

$$\frac{\partial^3}{\partial t^3} (t (A \cos (2t) + B \sin (2t))) + 4 \frac{\partial}{\partial t} (t (A \cos (2t) + B \sin (2t))) - \sin (2t)$$

$$(-8 B - 1) \sin (2t) - 8 A \cos (2t)$$

```
> solve({-8*B-1,-8*A},{A,B});
```

$$\{A = 0,\ B = \frac{-1}{8}\}$$

Hence, in this case we have the following general solution

```
> w(t):=a_1
> +a_2*cos(2*t)+a_3*sin(2*t)-1/8*t*sin(2*t);
```

$$w := t \mapsto a\_1 + a\_2 \cos (2t) + a\_3 \sin (2t) - 1/8\, t \sin (2t)$$

We set about satisfying the initial conditions

```
> equ1:=simplify(subs(t=0,w(t)));
```

$$equ1 := a\_1 + a\_2$$

```
> equ2:=simplify(subs(t=0,diff(w(t),t)));
```

$$equ2 := 2 a\_3$$

```
> equ3:=simplify(subs(t=0,diff(w(t),t$2))-1);
```

$$equ3 := -4 a\_2 - \frac{3}{2}$$

We solve for the coefficients in order that the initial conditions are satisfied.

```
> solve({equ1,equ2,equ3},{a_1
> ,a_2,a_3});
```

$$\{a\_3 = 0,\ a\_2 = \frac{-3}{8},\ a\_1 = \frac{3}{8}\}$$

Once again the same equation, but with the right-hand side $\exp 3t \sin(2t)$. This term is not included in the fundamental set of solutions for the non-homogeneous equation.

```
> de3:=diff(y(t),t$3) +
> 4*diff(y(t),t)-exp(3*t)*sin(2*t);
```

$$de3 := (\frac{\partial^3}{\partial t^3} \, y(t)) + 4 \, (\frac{\partial}{\partial t} \, y(t)) - e^{(3\,t)} \sin(2\,t)$$

```
>  simplify(subs(y(t)=exp(3*t)*(A*cos(2*t)+B*sin
>  (2*t)), de3));
```

$$\frac{\partial^3}{\partial t^3} \left(e^{3\,t}\left(A\cos\left(2\,t\right) + B\sin\left(2\,t\right)\right)\right) + 4\,\frac{\partial}{\partial t}\left(e^{3\,t}\left(A\cos\left(2\,t\right) + B\sin\left(2\,t\right)\right)\right) - e^{3\,t}\sin\left(2\,t\right)$$

$$3\,e^{3\,t}\left(\left(-18\,A + B - 1/3\right)\sin\left(2\,t\right) + \cos\left(2\,t\right)\left(A + 18\,B\right)\right)$$

```
>  solve({3*A+54*B, -54*A-1},{A,B});
```

$$\{A = \frac{1}{54}, B = \frac{-1}{972}\}$$

4.5.2   Variation of Parameters

In this session we show how we may use the variation of parameters method to solve the equation

$$\frac{d^3 y(t)}{dt^3} + \frac{d\,y(t)}{dt} = \tan(t).$$

First we must find a complete family of solutions. We do this by inputting the characteristic equation $r^3 + r = 0$, solve for the roots and write the solutions in terms of these.

```
>  de1:diff(y(t),t$3) + diff(y(t),t) -
>  tan(t);
```

$$(\frac{\partial^3}{\partial t^3} \, y(t)) + (\frac{\partial}{\partial t} \, y(t)) - \tan(t)$$

```
>  solve(r^3+r);
```

$$0, I, -I$$

Since the three roots are $\pm i$ and 0, we may take for real, independent solutions the set $\{1, \cos(t), \sin(t)\}$. hence, according to the variation of parameters method we seek a particular solution in the form

$$y_p(t) \; = \; v_1(t) + \cos(t)\,v_2(t) + \sin(t)\,v_3(t),$$

and seek to determine the unknown functions v_k, $k = 1..3$.

```
>  yp(t):=v1(t) + cos(t)*v2(t) +
>  sin(t)*v3(t);
```

$$yp := t \mapsto v1\,(t) + \cos\,(t)\,v2\,(t) + \sin\,(t)\,v3\,(t)$$

```
>  diff(yp(t),t);
```

$$(\frac{\partial}{\partial t} \text{v1}(t)) - \sin(t)\,\text{v2}(t) + \cos(t)\,(\frac{\partial}{\partial t}\,\text{v2}(t)) + \cos(t)\,\text{v3}(t) + \sin(t)\,(\frac{\partial}{\partial t}\,\text{v3}(t))$$

Since, the condition that (4.5.2) is satisfied is one condition and we need to determine three functions $v_k(t)$ we may impose two more conditions. These two conditions will be the equations listed as equ1 and equ2 here. If we differentiate (4.5.2) with respect to t we obtain

$$\frac{d\,y_p(t)}{dt} = \frac{d\,v_1(t)}{dt} + \cos(t)\frac{d\,v_2(t)}{dt} + \sin(t)\frac{d\,v_3(t)}{dt} - \sin(t)v_2(t) + \cos(t)v_3(t).$$

We set the sum of the first three term equal to zero and thereby get equation 1. Then we have

$$\frac{d\,y_p(t)}{dt} = -\,\sin(t)v_2(t) + \cos(t)v_3(t).$$

From this expression we compute

$$\frac{d^2 y_p(t)}{dt^2} = -\,\sin(t)\frac{d\,v_2(t)}{dt} + \cos(t)\frac{d\,v_3(t)}{dt} - \cos(t)v_2(t) - \sin(t)v_3(t).$$

By setting the sum of the first two terms equal to zero, we get the equation 2.

```
>  equ1:=(diff(v1(t),t))+cos(t)*diff(v2(t),t)+
```

```
sin(t)*diff(v3(t),t);
```

$$equ1 := (\frac{\partial}{\partial t}\,\text{v1}(t)) + \cos(t)\,(\frac{\partial}{\partial t}\,\text{v2}(t)) + \sin(t)\,(\frac{\partial}{\partial t}\,\text{v3}(t))$$

```
>  diff(-sin(t)*v2(t)+cos(t)*v3(t),t);
```

$$-\cos(t)\,\text{v2}(t) - \sin(t)\,(\frac{\partial}{\partial t}\,\text{v2}(t)) - \sin(t)\,\text{v3}(t) + \cos(t)\,(\frac{\partial}{\partial t}\,\text{v3}(t))$$

```
>  equ2:=-sin(t)*diff(v2(t),t)+cos(t)*diff(v3(t)
>  ,t);
```

$$equ2 := -\sin(t)\,(\frac{\partial}{\partial t}\,\text{v2}(t)) + \cos(t)\,(\frac{\partial}{\partial t}\,\text{v3}(t))$$

Equation 3 is obtained by substituting $y_p(t)$ and its derivatives into the differential equation.

```
>  equ3:=diff(-cos(t)*v2(t)-
```

```
sin(t)*v3(t),t)-sin( t)*v2(t)+cos(t)*v3(t)- tan(t);
```

$$equ3 := -\cos(t)\,(\frac{\partial}{\partial t}\,\text{v2}(t)) - \sin(t)\,(\frac{\partial}{\partial t}\,\text{v3}(t)) - \tan(t)$$

We now formally replace $\frac{d\,v_k}{dt}$ by $w_k(t)$ and solve the three equations for the $w_k(t)$.

```
>   simplify(solve({w1(t)+cos(t)*w2(t)+sin(t)*w3
>   (t),
```

```
-sin(t)*w2(t)+cos(t)*w3(t), -cos(t)*w2(t)-sin(t)*w3(t)-tan(t)},
    \{ w1(t),w2(t),w3(t)\}),trig);
```

$$\left\{ w1\,(t) = (-w3\,t + \cos(t))\sin(t)\,,\ w2\,(t) = -\sin(t)\,,\ w3\,(t) = -\frac{(\sin(t))^2}{\cos(t)} \right\}$$

$$-\frac{\sin(t)\left((t-1)(\cos(t))^2 - t\right)}{\cos(t)}$$

$$\frac{\sin(t)}{\cos(t)}$$

The $v_k(t)$ are obtained by individually integrating the $w_k(t)$.

```
>   int(sin(t)/cos(t),t);
```

$$-\ln(\cos(t))$$

```
>   int(-sin(t),t);
```

$$\cos(t)$$

```
>   int(1/cos(t)*(-sin(t)^2),t);
```

$$\sin(t) - \ln(\sec(t) + \tan(t))$$

```
>   with(linalg):
```

```
Warning, new definition for norm
```

```
Warning, new definition for trace
```

We now show a more efficient way to solve the above problem. We obtain this method by recognizing that the coefficient matrix of the linear equations for the *unknowns* $\frac{d\,v_1}{dt}$, $\frac{d\,v_2}{dt}$, and $\frac{d\,v_3}{dt}$ is the **Wronskian** matrix. Cramer's rule, which we discussed in an earlier session, provides a formula, using determinants, for the unknowns. Cramer's rule staes that if we want to solve the linear, algebraic equation $A\vec{x} = \vec{b}$, then the first unknown x_1 is found from the formula

$$x_1 = \begin{vmatrix} b_1 & a_{12} & a_{13} \\ b_2 & a_{22} & a_{23} \\ b_3 & a_{32} & a_{33} \end{vmatrix} \Big/ \begin{vmatrix} a_{11} & a_{12} & a_{13} \\ a_{21} & a_{22} & a_{23} \\ a_{31} & a_{32} & a_{33} \end{vmatrix}.$$

Similar formulas hold foe x_2 and x_3, where the second and third columns of the determinant of A is replaced by the column-vector $[b_1, b_2, b_3]$ respectively. First we use MAPLE to construct the Wronskian for the family of solutions $1, \cos(t), \sin(t)$. This is done using the command `wronskian`.

```
>   A:=vector([1,cos(t),sin(t)]);
```

$$A := [1, \cos(t), \sin(t)]$$

```
>   A0:=Wronskian(A,t);
```

$$A0 := \begin{bmatrix} 1 & \cos(t) & \sin(t) \\ 0 & -\sin(t) & \cos(t) \\ 0 & -\cos(t) & -\sin(t) \end{bmatrix}$$

```
>   simplify(det(A0),trig);
```

$$1$$

The various determinants which appear in the numerators can be constructed quite easily using the mouse to *mark* the determinant for A and then replace the particular column by the non-homogeneous term in the linear, algebraic system.

```
>   A1:simplify(det([[0, cos(t), sin(t)], [0,
>   -sin(t), cos(t)], [tan(t), -cos(t), -sin(t)]]),trig);
```

$$\frac{\sin(t)}{\cos(t)}$$

```
>   A2:=det([[1, 0, sin(t)], [0, 0, cos(t)], [0,
>   tan(t), -sin(t)]]);
```

$$A2 := -\cos(t)\tan(t)$$

```
>   A3:=det([[1, cos(t), 0], [0, -sin(t), 0], [0,
>   -cos(t), tan(t)]]);
```

$$A3 := -\sin(t)\tan(t)$$

```
>   v1(t):=int(sin(t)/cos(t),t);
```

$$v1(t) := -\ln(\cos(t))$$

$$v1 := t \mapsto -\ln(\cos(t))$$

```
>   v2(t):=int(-cos(t)*tan(t),t);
```

$$v2 := t \mapsto \cos(t)$$

```
>   v3(t):=int(-sin(t)*tan(t),t);
```

$$v3 := t \mapsto \sin(t) - \ln(\sec(t) + \tan(t))$$

4.5.3 Further Remarks on the Variation of Parameters Method

```
>  with(linalg):
```

Warning, new definition for norm

Warning, new definition for trace

An alternate MAPLE approach, which depends on the third-order Wronskian matrix is given below.

```
>  Wron:=(y1,y2,y3)->linalg[matrix](3,3,[y1,y2,y3,
>  diff(y1,x),diff(y2,x),diff(y3,x),diff(y1,x$2),diff(y2,x$2),
>  diff(y3,x$2)]);
```

$$Wron := (y1,\ y2,\ y3) \to$$

$$linalg_{matrix}(3,\ 3,\ [y1,\ y2,\ y3,\ \tfrac{\partial}{\partial x}\,y1,\ \tfrac{\partial}{\partial x}\,y2,\ \tfrac{\partial}{\partial x}\,y3,\ \tfrac{\partial^2}{\partial x^2}\,y1,\ \tfrac{\partial^2}{\partial x^2}\,y2,\ \tfrac{\partial^2}{\partial x^2}\,y3])$$

Let us illustrate our method by applying it to the third-order equation depending on the parameter α

```
>  deq1:=diff(y(x),x$3)-alpha*diff(y(x),x$2)+diff(y(x),x)-alpha*y(x);
```

$$deq1 := (\tfrac{\partial^3}{\partial x^3}\,\mathrm{y}(x)) - \alpha\,(\tfrac{\partial^2}{\partial x^2}\,\mathrm{y}(x)) + (\tfrac{\partial}{\partial x}\,\mathrm{y}(x)) - \alpha\,\mathrm{y}(x)$$

The characteristic roots are clearly $\pm i, \alpha$. This tells us that a fundamental system of solutions is given by $\sin(x)$, $\cos(x)$, $\exp \alpha x$. We check that these solutions really form a fundamental system by computing their Wronskian.

```
>  Wron(sin(x),cos(x),exp(alpha*x));
```

$$\begin{bmatrix} \sin(x) & \cos(x) & e^{(\alpha\,x)} \\ \cos(x) & -\sin(x) & \alpha\,e^{(\alpha\,x)} \\ -\sin(x) & -\cos(x) & \alpha^2\,e^{(\alpha\,x)} \end{bmatrix}$$

```
>  simplify(Determinant(%),trig);
```

$$-e^{\alpha\,x}\left(\alpha^2 + 1\right)$$

This computation can be automated by the MAPLE function `wronsk`.

```
>  wronsk :=(y1,y2,y3)->simplify(
>  det(Wron(y1,y2,y3)));
```

$$wronsk := (y1,\ y2,\ y3) \to \mathrm{simplify}(\mathrm{Determinant}(\mathrm{Wron}(y1,\ y2,\ y3)))$$

```
>  wronsk(sin(x),cos(x),exp(alpha*x));
```

$$-e^{\alpha\,x}\left(\alpha^2 + 1\right)$$

Let us now construct the general solution of the non-homogeneous equation

```
>  deq2:=deq1-tan (x);
```

$$deq2 := (\frac{\partial^3}{\partial x^3} \, \mathrm{y}(x)) - \alpha \, (\frac{\partial^2}{\partial x^2} \, \mathrm{y}(x)) + (\frac{\partial}{\partial x} \, \mathrm{y}(x)) - \alpha \, \mathrm{y}(x) - \tan(x)$$

We use `dsolve` to solve this differential equation, but first we need to load Detools.

```
> with(DEtools):
```

Warning, the name adjoint has been redefined

```
> dsolve(deq2,y(x));
```

$$y(x) = -\frac{(\tan)(\alpha x + 1)}{\alpha^2} + \_C1 \, \cos(x) + \_C2 \, \sin(x) + \_C3 \, e^{\alpha x}$$

Let us check MAPLE 's result using the method of variation of parameters. We recall that we need to solve for the functions $v_k(x)$, $k = 1, 2, 3$ which act as coefficients of the fundamental solutions in the representation of the particular solution. Let us formulate our ideas in general, and assume that the fundamental system of solutions for a third-order differential equation is given by $y_1(x)$, $y_2(x)$, $y_3(x)$. The particular solution is then represented as

$$y_p(x) = v_1(x)y_1(x) + v_2(x)y_2(x) + v_3(x)y_3(x).$$

We need to compute the coefficient functions $v_k(x)$, $k = 1, 2, 3$, and these satisfy the system

$$
\begin{array}{l}
\frac{d\,v_1}{dx}y_1(x) + \frac{d\,v_2}{dx}y_2(x) + \frac{d\,v_3}{dx}y_3(x) = 0 \\
\frac{d\,v_1}{dx}\frac{d\,y_1(x)}{dx} + \frac{d\,v_2}{dx}\frac{d\,y_2(x)}{dx} + \frac{d\,v_3}{dx}\frac{d\,y_3(x)}{dx} = 0 \\
\frac{d\,v_1}{dx}\frac{d^2\,y_1(x)}{dx^2} + \frac{d\,v_2}{dx}\frac{d^2\,y_2(x)}{dx^2} + \frac{d\,v_3}{dx}\frac{d^2\,y_3(x)}{dx^2} = 0.
\end{array}
$$

This system of equations may be solved for the $\frac{d\,v_k}{dx}$ using Cramer's rule. We input this result as a MAPLE function, where the functions $u_k(x) := \frac{d\,v_k(x)}{dx}$.

```
> u1:=(g,y1,y2,y3) ->
> int(det(linalg[matrix](3,3,[0,y2,y3,0,diff(y2,x),
> diff(y3,x),g,diff(y2,x$2),
> diff(y3,x$2)]))/wronsk(y1,y2,y3),x):
> u1(tan(x),sin(x),cos(x),exp(alpha*x));
```

$$\frac{\cos(x)\,\alpha}{\alpha^2 + 1} + \frac{\sin(x)}{\alpha^2 + 1} - \frac{\ln(\sec(x) + \tan(x))}{\alpha^2 + 1}$$

```
> u2:=(g,y1,y2,y3) ->
> -int(det(linalg[matrix](3,3,[0,y1,y3,
> 0,diff(y1,x),diff(y3,x),g,diff(y1,x$2),
> diff(y3,x$2)]))/wronsk(y1,y2,y3),x):
> u2(tan(x),sin(x),cos(x),exp(alpha*x));
```

$$-\frac{\sin(x)\,\alpha}{\alpha^2 + 1} + \frac{\alpha \, \ln(\sec(x) + \tan(x))}{\alpha^2 + 1} + \frac{\cos(x)}{\alpha^2 + 1}$$

```
):
>   u3:=(g,y1,y2,y3) ->
>   int(det(linalg[matrix](3,3,[0,y2,y1,0,
>   diff(y2,x),diff(y1,x),g,diff(y2,x$2),diff(y1,x$2)]))
>     /wronsk(y1,y2,y3),x):
>   u3(tan(x),sin(x),cos(x),exp(alpha*x));
```

$$-\frac{\displaystyle\int \frac{\sin(x)\,e^{(-\alpha\,x)}}{\cos(x)}\,dx}{\alpha^2+1}$$

Let us try our MAPLE function, based on the method of variation of parameters, on the differential equation we solved using dsolve.

```
>   first(x):=simplify(u1(tan(x),sin(x),cos(x),exp(alpha*x)));
```

$$\text{first}(x) := \frac{\alpha\cos(x)}{\alpha^2+1} + \frac{\sin(x)}{\alpha^2+1} - \frac{\ln(\sec(x)+\tan(x))}{\alpha^2+1}$$

```
>   second(x):=simplify(u2(tan(x),sin(x),cos(x),e
>   xp(alpha*x)));
```

$$\text{second}(x) := -\frac{\alpha\sin(x) - \alpha\ln(\dfrac{1+\sin(x)}{\cos(x)}) - \cos(x)}{\alpha^2+1}$$

```
>   third(x):=simplify(u3(tan(x),sin(x),cos(x),ex
>   p(alpha*x)));
```

$$\text{third}(x) := -\frac{\displaystyle\int \frac{\sin(x)\,e^{(-\alpha\,x)}}{\cos(x)}\,dx}{\alpha^2+1}$$

Let us automate the Cramer's rule using MAPLE. To this end we input the three equations one gets by substituting in the form of the particular solution in for $y_p(x)$, and its derivatives up second order, plus the differential equation. Equation (1) is first *arbitrary* condition we impose.

```
>   eq1:=(y1,y2,y3)->u1_prime*y1+u2_prime*y2+u3_p
>   rime*y3;
```

$$eq1 := (y1,\,y2,\,y3) \to u1\_prime\,y1 + u2\_prime\,y2 + u3\_prime\,y3$$

Equation (2) is the second *arbitrary condition* we impose.

```
>   eq2:=(y1,y2,y3)->u1_prime*diff(y1,x)+u2_prime*diff(y2,x)
>   +u3_prime*diff(y3,x);
```

$$eq2 := (y1,\,y2,\,y3) \to u1\_prime\,(\tfrac{\partial}{\partial x}\,y1) + u2\_prime\,(\tfrac{\partial}{\partial x}\,y2) + u3\_prime\,(\tfrac{\partial}{\partial x}\,y3)$$

Equation (3) is the condition the function must satisfy the non-homogeneous differential equation.

```
>   eq3:=(y1,y2,y3)->u1_prime*diff(y1,x$2)+u2_pri
>   me*diff(y2,x$2)+
>   u3_prime*diff(y3,x$2)-g;
```

$eq3 := (y1,\, y2,\, y3) \rightarrow$
    $u1\_prime \operatorname{diff}(y1,\, x\,\$\,2) + u2\_prime \operatorname{diff}(y2,\, x\,\$\,2) + u3\_prime \operatorname{diff}(y3,\, x\,\$\,2) - g$

```
>   e1:=eq1(sin(x),cos(x),exp(x));
```

$$e1 := u1\_prime\,\sin(x) + u2\_prime\,\cos(x) + u3\_prime\,e^{x}$$

```
>   e2:=eq2(sin(x),cos(x),exp(x));
```

$$e2 := u1\_prime\,\cos(x) - u2\_prime\,\sin(x) + u3\_prime\,e^{x}$$

```
>   e3:=eq3(sin(x),cos(x),exp(x));
```

$$e3 := -u1\_prime\,\sin(x) - u2\_prime\,\cos(x) + u3\_prime\,e^{x} - g$$

We now solve for the derivatives of the coefficients of the $y_k(x)$, called here $u'_k(x)$.

```
>   solve({e1,e2,e3},{u1_prime,u2_prime,u3_pri
>   me});
```

$$\left\{ u1\_prime = -\frac{g\,(\cos(x) + \sin(x))}{2\,(\cos(x))^{2} + 2\,(\sin(x))^{2}}, \right.$$

$$u2\_prime = -\frac{(\cos(x) - \sin(x))\,g}{2\,(\cos(x))^{2} + 2\,(\sin(x))^{2}},\ u3\_prime = 1/2\,\frac{g}{e^{x}} \right\}$$

$$\{u1\_prime = -1/2\,g\,(\cos(x) + \sin(x)),\ u2\_prime$$
$$= -1/2\,(\cos(x) - \sin(x))\,g,\ u3\_prime = 1/2\,e^{-x}g\}$$

Chapter 5

Applications of Second-Order Linear Equations

5.1 Simple Harmonic Motion

The purpose of this chapter is to study the simple harmonic motion as an application of second-order linear differential equations with constant coefficients. Mechanical vibrations can be modelled by solving the following IVP:

$$m\frac{d^2}{dt^2}u + \gamma\frac{d}{dt}u + ku = F(t), \quad t > 0$$

$$u(0) = u_0 \quad \text{and} \quad \frac{d}{dt}u(0) = u_1,$$

,where $m > 0, \gamma \geq 0$ and $k \geq 0$ are constants. Physically m is the mass of the bob, k the spring coefficient, and γ the damping coefficient. For free vibration, the driving force term $F(t)$ is equal to zero. Then there are two cases: *undamped free vibration*, in which case $\gamma = 0$, and *damped free vibration* while $\gamma \neq 0$. Similarly, for the forced vibrations in which case $F \neq 0$, there are undamped and damped, forced vibrations according to whether γ equals to zero or not.

In the following, we consider some examples and show how one may use Maple to solve the problems. In these examples, some of the relevant Maple commands in connection with second-order differential equations will be introduced. These are `coeff`, `combine`, `collect`, and `map`.

5.2 General Solutions

We begin with the basics for the second-order homogeneous linear equations with constant coefficients.

```
>  Ly:=a*diff(y(x),x$2)+b*diff(y(x),x)+c*y(x);
```

$$Ly := a\left(\frac{\partial^2}{\partial x^2}\,y(x)\right) + b\left(\frac{\partial}{\partial x}\,y(x)\right) + c\,y(x)$$

```
>  deq1:=subs({a=1,b=4,c=3},Ly)=0;
```

$$deq1 := \left(\frac{\partial^2}{\partial x^2}\,y(x)\right) + 4\left(\frac{\partial}{\partial x}\,y(x)\right) + 3\,y(x) = 0$$

```
>  with(DEtools):
>  gsoln1:=dsolve(deq1,y(x));
```

$$gsoln1 := y(x) = \_C1\,e^{(-x)} + \_C2\,e^{(-3\,x)}$$

```
>  ics:=y(0)=1,D(y)(0)=0;
```

$$ics := y(0) = 1,\ D(y)(0) = 0$$

```
>  soln1:=dsolve({deq1,ics},y(x));
```

$$soln1 := y(x) = -1/2\,e^{-3\,x} + 3/2\,e^{-x}$$

```
>  soln1:=expand(%);
```

$$soln1 := y(x) = -1/2\,(e^x)^{-3} + 3/2\,(e^x)^{-1}$$

The following computation shows how to combine the trigonometric terms into a single term by using the command combine:

```
>  soln2:=dsolve({lhs(deq1)=cos(x),ics},y(x));
```

$$soln2 := y(x) = -\frac{7\,e^{-3\,x}}{20} + 5/4\,e^{-x} + 1/10\,\cos(x) + 1/5\,\sin(x)$$

```
>  expand(%);
```

$$y(x) = -\frac{7}{20\,(e^x)^3} + 5/4\,(e^x)^{-1} + 1/10\,\cos(x) + 1/5\,\sin(x)$$

$$Yc := \frac{1}{10}\,\cos(x) + \frac{1}{5}\,\sin(x)$$

```
>  a=1/10: b=1/5:
>  Yc:=sqrt(a^2+b^2)*(cos(phi)*cos(x)+sin(phi)*s
>  in(x));
```

$$Yc := \sqrt{a^2 + b^2}\,(\cos(\varphi)\cos(x) + \sin(\varphi)\sin(x))$$

```
>  combine(%);
```

$$\sqrt{a^2 + b^2}\,\cos(x - \varphi)$$

One may use the command numeric to obtain the numerical solutions:

```
>  nsol1:=dsolve({deq1,ics},y(x),numeric);
```

$$nsol1 := \mathbf{proc}(rkf45\_x) \ldots \mathbf{end}$$

```
>   nsol1(2);
```

$$[x = 2, \text{y}(x) = 0.2017635497397481, \frac{\partial}{\partial x} \text{y}(x) = -0.1992848000035134]$$

```
>   y1(2):=evalf(subs(x=2,rhs(soln1)));
```

$$\text{y1}(2) := .2017635487$$

5.3   Method of Undetermined Coefficients

This solution is only valid for x up to π. The problem is that the right hand side is contained in the general solution of the corresponding homogeneous equation. Note:

```
>   with(DEtools):
>   deq3:=diff(y(x),x$2)+y(x)=1/2*cos(x);
```

$$deq3 := \frac{d^2}{dx^2} y(x) + y(x) = 1/2 \cos(x)$$

```
>   Yc:=dsolve(lhs(deq3)=0,y(x));
```

$$Yc := y(x) = \_C1 \sin(x) + \_C2 \cos(x)$$

Now let us solve it as if we do by hand. First we look for a particular solution by the method of undetermined coefficients in the form:

```
>   Yp:=x*(A*cos(x)+B*sin(x));
```

$$Yp := x(A\cos(x) + B\sin(x))$$

```
>   subs(y(x)=Yp,deq3);
```

$$(\frac{\partial^2}{\partial x^2} x(A\cos(x) + B\sin(x))) + x(A\cos(x) + B\sin(x)) = \frac{1}{2}\cos(x)$$

```
>   simplify(%);
```

$$-2A\sin(x) + 2B\cos(x) = 1/2\cos(x)$$

```
>   eq:=lhs(%)-rhs(%)=0;
```

$$eq := -2\,A\sin(x) + 2\,B\cos(x) - \frac{1}{2}\cos(x) = 0$$

```
>  solve({coeff(lhs(eq),cos(x))=0,coeff(lhs(eq)
>  ,sin(x))=0},{A,B});
```

$$\{A = 0,\ B = \frac{1}{4}\}$$

```
>  assign(%):
>  gsoln3:=rhs(Yc)+Yp;
```

$$gsoln3 := \_C1\cos(x) + \_C2\sin(x) + \frac{1}{4}x\sin(x)$$

```
>  subs(x=0,gsoln3)=0;
```

$$\_C1\sin(0) + \_C2\cos(0) = 0$$

```
>  eq1:=simplify(%);
```

$$eq1 := \_C2 = 0$$

```
>  subs(x=0,diff(gsoln3,x))=0;
```

$$-\_C1\sin(0) + \_C2\cos(0) + \frac{1}{4}\sin(0) = 0$$

```
>  eq2:=simplify(%);
```

$$eq2 := \_C1 = 0$$

```
>  solve({eq1,eq2},{_C1,_C2});
```

$$\{\_C1 = 0,\ \_C2 = 0\}$$

```
>  mysoln3:=subs(%,gsoln3);
```

$$mysoln3 := 1/4\,x\sin(x)$$

```
>  soln3:=dsolve({deq3,y(0)=0,D(y)(0)=0},y(x))
>  ;
```

$$soln3 := y(x) = 1/4\,x\sin(x)$$

5.4   Additional Useful Commands

Some additional commands such as map and collect should be handy for dealing with long expressions as follows:

```
>  with(DEtools):
>  macro(w=omega, W=Omega, c=gamma);
```

$$w, W, c$$

```
>  -(a*cos(w*t)*w^2*m+b*sin(b*t)*w^2*m+c*w*a*sin(w*t)
>  -c*w*b*cos(w*t)-W^2
>  *m*a*cos(w*t)-W^2*m*b*sin(w*t))/m=F0*sin(w*t)/m;
```

$$a\cos{(\omega\, t)}\,\omega^2 m + b\sin{(bt)}\,\omega^2 m + \gamma\,\omega\, a\sin{(\omega\, t)}$$
$$-\frac{-\gamma\,\omega\, b\cos{(\omega\, t)} - \Omega^2 ma\cos{(\omega\, t)} - \Omega^2 mb\sin{(\omega\, t)}}{m} = \frac{F0\,\sin{(\omega\, t)}}{m}$$

```
>  eq:=lhs(%)-rhs(%)=0;
```

$$eq := -(a\cos(\omega\, t)\,\omega^2\, m + b\sin(bt)\,\omega^2\, m + \gamma\,\omega\, a\sin(\omega\, t) - \gamma\,\omega\, b\cos(\omega\, t)$$
$$- \Omega^2\, m\, a\cos(\omega\, t) - \Omega^2\, m\, b\sin(\omega\, t))/m - \frac{F0\,\sin(\omega\, t)}{m} = 0$$

```
>  col:=map(collect,eq,[cos(w*t),sin(w*t)]);
```

$$col := -\frac{(a\,\omega^2\, m - \gamma\,\omega\, b - \Omega^2\, m\, a)\cos(\omega\, t)}{m} + (-\frac{\gamma\,\omega\, a - \Omega^2\, m\, b}{m} - \frac{F0}{m})\sin(\omega\, t)$$
$$- b\sin(bt)\,\omega^2 = 0$$

```
>  {coeff(lhs(col),cos(w*t))=0,coeff(lhs(col),sin(w*t))};
```

$$\left\{ -\frac{-\Omega^2 mb + \gamma\,\omega\, a}{m} - \frac{F0}{m}, -\frac{-\Omega^2 ma + am\omega^2 - \gamma\,\omega\, b}{m} = 0 \right\}$$

```
>  eq1:=map(collect,%,[a,b]);
```

$$eq1 := \{-\frac{\gamma\,\omega\, a}{m} + \Omega^2 b - \frac{F0}{m}, -\frac{(\omega^2\, m - \Omega^2\, m)\, a}{m} + \frac{\gamma\,\omega\, b}{m} = 0\}$$

```
>  solve(eq1,{a,b});
```

$$\{a = -\frac{F0\,\gamma\,\omega}{\gamma^2\,\omega^2 - \Omega^2\, m^2\,\omega^2 + \Omega^4\, m^2},\ b = \frac{m\,(-\omega^2 + \Omega^2)\, F0}{\gamma^2\,\omega^2 - \Omega^2\, m^2\,\omega^2 + \Omega^4\, m^2}\}$$

Finally, for trigonometric simplification, one may find the maple procedure trigsubs useful. However, before one issues the command, the procedure should first be loaded from the library:

```
>  readlib(trigsubs);
```

$$\mathbf{proc}(s, f)\, \ldots\, \mathbf{end\ proc}$$

```
>  sin(theta)+sin(phi):%=trigsubs(%);
```

$$\sin{(\theta)} + \sin{(\varphi)} = [2\,\sin{(\theta/2 + \varphi/2)}\cos{(\theta/2 - \varphi/2)}]$$

```
>  lhs(%)+op(1,rhs(%));
```

$$\sin{(\theta)} + \sin{(\varphi)} + 2\,\sin{(\theta/2 + \varphi/2)}\cos{(-\theta/2 + \varphi/2)}$$

5.5   Computer Lab

In the IVP, let $\omega_0^2 := k/m$. Then ω_0 is called the *natural frequency*. Suppose that the driving force $F(t)$ is a sinusoid in the form:

$$F(t) = F_0 cos(\omega t),$$

where $F_0 > 0$ is the *amplitude* and ω is the *driving frequency*. Suppose that there is **no damping**, i.e., $\gamma = 0$.

Beats: Find the solution of the IVP that satisfies the homogeneous initial conditions (i.e., $u_0 = 0$ and $u_1 = 0$) by assuming that $\omega \neq \omega_0$. Show that the solution can be rewritten in the form:

$$u(t) = [\frac{2F_0}{m(\omega_0^2 - \omega^2)} sin(\frac{\omega_0 - \omega}{2}t)]sin(\frac{\omega_0 + \omega}{2}t).$$

Note that if ω is close to ω_0, the difference $\omega - \omega_0$ is small, so that the period of the sin function, $sin(\frac{\omega_0 - \omega}{2}t)$ is large and the solution displays the phenomenon of *beats*. This is what musicians are listening to when they turn their instruments.

Plot the solution by taking $\omega_0^2 = 1, \omega = 0.8, F_0/m = 0.5$ (see p.185 in the text).

Resonance: Suppose now $\omega = \omega_0$. Find the solution of the IVP that satisfies the homogeneous initial conditions. Show that the solution is *sinusoid* with an unbounded amplitude depending on t. This phenomenon is called *resonance*.

Plot the solution as a function t for $\omega_0 = 1, \omega = 1, F_0/m = 0.5$ (see p. 185 in the text).

5.6   Supplementary Maple Programs

5.6.1   The Phenomenon of Beats

In this section we investigate the phenomenon of beats in a mechanical system.

```
>   with(DEtools):
```

We begin by entering our macros.

```
>   macro(w=omega, W=Omega); assume(w>W);
```

$$w, W$$

`>  assume(Fo>0,m>0);`

We consider the differential equation where the forcing term frequency is not the same as the natural frequency of the homogeneous equation.

`>  deq:=diff(u(t),t$2)+W^2*u(t)=Fo/m*cos(w*t);`

$$deq := \frac{d^2}{dt^2}u(t) + \omega^2 u(t) = \frac{F0 \, \cos(\omega t)}{m}$$

To simplify the calculations we assume homogeneous initial conditions.

`>  ics:=u(0)=0,D(u)(0)=0;`

$$ics := u(0) = 0, \, D(u)(0) = 0$$

dsolve then provides us with a concise solution.

`>  dsolve({deq,ics},u(t));`

$$u(t) = -\frac{\cos\left(\sqrt{\Omega^2}t\right) Fo}{m\left(\Omega^2 - \omega^2\right)} + \frac{Fo \, \cos(\omega t)}{m\left(\Omega^2 - \omega^2\right)}$$

`>  simplify(%);`

$$u(t) = -\frac{Fo\left(\cos(t|\Omega|) - \cos(\omega t)\right)}{m\left(\Omega^2 - \omega^2\right)}$$

`>  soln:=rhs(%);`

$$soln := -\frac{Fo\left(\cos(t|\Omega|) - \cos(\omega t)\right)}{m\left(\Omega^2 - \omega^2\right)}$$

In order to express the sum of the two cosines as a product of two sine functions we make use `trigsubs` which we learned about in the last section.

`>  readlib(trigsubs);`

$$\mathbf{proc}(s, f) \dots \mathbf{end \ proc}$$

`>  soln:=Fo*trigsubs(-cos(t*abs(Omega))+cos(omega*t))/1/m/`
`>  (Omega^2-omega^2);`

$$soln := \frac{Fo\left[-2\sin(1/2\,\omega t + 1/2\,t|\Omega|)\sin(1/2\,\omega t - 1/2\,t|\Omega|)\right]}{m\left(\Omega^2 - \omega^2\right)}$$

If the two frequencies ω and Ω are close but not equal we observe the phenomenon of **beats**. We choose these parameters accordingly. The beats can be seen in the following plot.

`>  soln:=simplify(subs({Fo=1,w=10.5,W=10,m=1},soln));`

$$soln := [.1951219512 \sin(10.25000000\,t) \sin(.2500000000\,t)]$$

`>  with(plots):`
`>  plot(soln,t=0..50);`

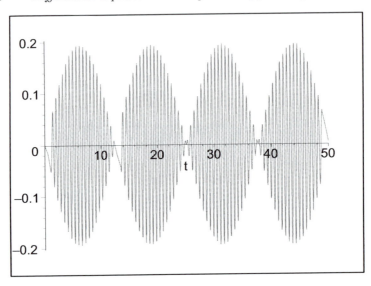

5.6.2   The Phenomenon of Resonance

```
>   with(DEtools):
>   macro(w=omega);   assume(w>0,Fo>0,m>0);
```

$$w$$

```
>   deq:=diff(u(t),t$2) +w^2*u(t)=F0/m*cos(w*t);
```

$$deq := \frac{\mathrm{d}^2}{\mathrm{d}t^2}u\left(t\right) + \omega^2 u\left(t\right) = \frac{F0\ \cos\left(\omega\,t\right)}{m}$$

```
>   ics:=u(0)=0,D(u)(0)=0;
```

$$ics := u(0) = 0,\ \mathrm{D}(u)(0) = 0$$

```
>   dsolve({deq,ics},u(t));
```

$$u\left(t\right) = 1/2\,\frac{F0\ \sin\left(\omega\,t\right)t}{\omega\,m}$$

```
>   soln:=simplify(subs({w=1,F0=0.5*m},%));
```

$$soln := u(t) = 0.2500000000\sin(t)\,t$$

```
>   with(plots):
```

Warning, the name changecoords has been redefined

```
>   p1:=plot(rhs(soln),t=0..100):
>   U1:=0.25*t;
```

$$U1 := 0.25\,t$$

```
>   p2:=plot({U1,-U1},t=0..100,color=RED): display(p1,p2);
```

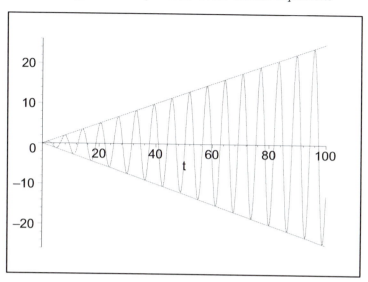

5.7   Particular Solutions

In the previous chapter we showed how to find particular solutions using the method of undetermined coefficients. However, we may obtain the general solution also by using dsolve. We can get a particular solution easily by setting the general constants equal to zero.

```
>   with(DEtools):
>   macro(W=Omega,w=omega,c=gamma);
```

$$w, W, c$$

```
>   deq:=diff(u(t),t$2)+c/m*diff(u(t),t)+W^2*u(t)=Fo/m*sin(w*t);
```

$$deq := \frac{d^2}{dt^2}u(t) + \frac{\gamma \frac{d}{dt}u(t)}{m} + \Omega^2 u(t) = \frac{Fo\,\sin(\omega\,t)}{m}$$

```
>   soln:=dsolve(deq,u(t));
```

$$soln := u(t) = e^{(-1/2\,\frac{(\gamma - \sqrt{\gamma^2 - 4\,\Omega^2\,m^{\sim 2}})\,t}{m^{\sim}})}\,\_C2 + e^{(-1/2\,\frac{(\gamma + \sqrt{\gamma^2 - 4\,\Omega^2\,m^{\sim 2}})\,t}{m^{\sim}})}\,\_C1$$
$$-\frac{\left((-m^{\sim}\Omega^2 + \omega^{\sim 2}\,m^{\sim})\sin(\omega^{\sim}\,t) + \omega^{\sim}\cos(\omega^{\sim}\,t)\,\gamma\right)Fo^{\sim}}{\omega^{\sim 4}\,m^{\sim 2} + (\gamma^2 - 2\,\Omega^2\,m^{\sim 2})\,\omega^{\sim 2} + \Omega^4\,m^{\sim 2}}$$

```
>   Up:=subs(_C1=1,_C2=0,rhs(soln));
```

$$Up := e^{-1/2\,\frac{\left(\gamma + \sqrt{-4\,\Omega^2 m^2 + \gamma^2}\right)t}{m}} - \frac{Fo\left(-\sin(\omega\,t)\,\Omega^2 m + \sin(\omega\,t)\,m\omega^2 + \gamma\,\cos(\omega\,t)\,\omega\right)}{m^2\omega^4 - 2\,\Omega^2 m^2\omega^2 + \Omega^4 m^2 \quad + \gamma^2\omega^2}$$

```
>  Up:=convert(Up,trig);
```

$$Up := \cosh\left(1/2\,\frac{\left(\gamma+\sqrt{-4\,\Omega^2 m^2+\gamma^2}\,\right)t}{m}\right) - \sinh\left(1/2\,\frac{\left(\gamma+\sqrt{-4\,\Omega^2 m^2+\gamma^2}\,\right)t}{m}\right)$$

$$-\frac{Fo\left(-\sin\left(\omega\,t\right)\Omega^2 m + \sin\left(\omega\,t\right)m\omega^2 + \gamma\,\cos\left(\omega\,t\right)\omega\right)}{m^2\omega^4 - 2\,\Omega^2 m^2\omega^2 + \Omega^4 m^2 \qquad + \gamma^2\omega^2}$$

```
>  Up:=Re(subs({W=10,w=10.1,Fo=1,m=1,c=0.1},Up));
```

$$Up := \operatorname{Re}\left(\cosh((0.05000000000 + 9.999875000\,I)\,t)\right.$$
$$- \sinh((0.05000000000 + 9.999875000\,I)\,t) - 0.3972175013\sin(10.1\,t)$$
$$\left.- 0.1995968539\cos(10.1\,t)\right)$$

```
>  with(plots):
>  plot(Up,t=0..10);
```

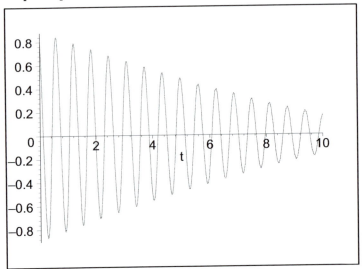

5.8   Computer Lab

In the IVP, let $\omega_0^2 := k/m$. Then ω_0 is called the *natural frequency*. Suppose that the driving force $F(t)$ is a sinusoid in the form:

$$F(t) = F_0 cos(\omega t),$$

where $F_0 > 0$ is the *amplitude* and ω is the *driving frequency*. Suppose that the motion is **damped** with homogeneous initial conditions, i.e., $\gamma \neq 0, u_0 = 0, u_1 = 0$ in IVP.

- Find a particular solution $u_p(t)$ of $L[u] = F(t)$, and show that the solution can be rewritten in the form:

$$u_p(t) = R(\omega)sin(\omega t - \delta),$$

where

$$R(\omega) := \frac{F_0}{\sqrt{m^2(\omega_0^2 - \omega^2)^2 + \gamma^2\omega^2}}, \quad \delta := tan^{-1}\frac{\gamma\omega}{m(\omega_0^2 - \omega^2)}.$$

The dimensionless parameter δ is called the *phase* or *phase angle*.

- Show that the maximum amplitude $R(\omega)$ occurs when

$$\omega^2 = \omega_0^2(1 - \frac{\gamma^2}{2km}).$$

- Rewrite $R(\omega)$ in the form:

$$R(\omega) = \frac{F_0}{m\omega_0^2} A(\frac{\omega}{\omega_0}, \Gamma),$$

where

$$A(\frac{\omega}{\omega_0}, \Gamma) := \frac{1}{\sqrt{(1 - (\frac{\omega}{\omega_0})^2)^2 + \Gamma(\frac{\omega}{\omega_0})^2}}, \quad \Gamma := \frac{\gamma^2}{m^2\omega_0^2}.$$

Plot $A(x, \Gamma)$ as a function of $x = \frac{\omega}{\omega_0}$ for $\Gamma = 0.2, 0.5, 1.0, 2.0,$ and 3.0 (see p. 187 in the text).

- Solve the IVP with $\gamma/m = 2, \omega_0^2 = 1, \omega = 0.3, F_0/m = 3$, and plot the solution and $\frac{1}{k} \times$ the forcing term as a function of t in the same figure.

- The solution of the previous problem is eventually dominated by the particular solution as time increases. Verify this by taking the limit of the solution as $t \to \infty$.

5.9   Supplementary Maple Programs

5.9.1   Resonance Curves

```
>   with(DEtools):
>   macro(W=Omega,w=omega,c=gamma);
>   W,w,c;
```

$$\Omega, \omega, \gamma$$

```
>   assume(F0>0,m>0);
>   deq:=diff(u(t),t$2)+c/m*diff(u(t),t)+W^2*u(t)
>   =F0/m*sin(w*t);
```

$$deq := (\frac{\partial^2}{\partial t^2}\,u(t)) + \gamma\,(\frac{\partial}{\partial t}\,u(t)) + \Omega^2\,u(t) = F0^{\tilde{}}\sin(\omega\,t)$$

```
>   dsolve(deq,u(t));
```

$$u(t) = \frac{F0^{\tilde{}}\,(-\omega\cos(\omega\,t)\,\gamma + \sin(\omega\,t)\,\Omega^2 - \sin(\omega\,t)\,\omega^2)}{\gamma^2\,\omega^2 + \Omega^4 - 2\,\Omega^2\,\omega^2 + \omega^4}$$
$$+\ \_C1\,e^{(1/2(-\gamma+\sqrt{-(2\Omega-\gamma)(2\Omega+\gamma)})\,t)} + \_C2\,e^{(-1/2(\gamma+\sqrt{-(2\Omega-\gamma)(2\Omega+\gamma)})\,t)}$$

```
>   subs(_C1=0,_C2=0,%);
```

$$u(t) = -\frac{F0\,\left(-\sin(\omega\,t)\,\Omega^2 m + \sin(\omega\,t)\,m\omega^2 + \cos(\omega\,t)\,\gamma\,\omega\right)}{\Omega^4 m^2 - 2\,\Omega^2 m^2\omega^2 + m^2\omega^4 + \gamma^2\omega^2}$$

```
>   Up:=rhs(map(collect,%,[cos(w*t),sin(w*t)]));
```

$$Up := -\frac{F0\,\gamma\,\omega\,\cos(\omega\,t)}{\Omega^4 m^2 - 2\,\Omega^2 m^2\omega^2 + m^2\omega^4 + \gamma^2\omega^2} - \frac{F0\,\left(-\Omega^2 m + m\omega^2\right)\sin(\omega\,t)}{\Omega^4 m^2 - 2\,\Omega^2 m^2\omega^2 + m^2\omega^4 + \gamma^2\omega^2}$$

```
>   A:=coeff(Up,cos(w*t));
>   B:=coeff(Up,sin(w*t));
```

$$A := -\frac{F0\,\gamma\,\omega}{\Omega^4 m^2 - 2\,\Omega^2 m^2\omega^2 + m^2\omega^4 + \gamma^2\omega^2}$$

$$B := -\frac{F0\,\left(-\Omega^2 m + m\omega^2\right)}{\Omega^4 m^2 - 2\,\Omega^2 m^2\omega^2 + m^2\omega^4 + \gamma^2\omega^2}$$

```
>   R(w):=sqrt(A^2+B^2);
```

$$R(\omega) := \sqrt{\frac{F0^{\tilde{}2}\,\omega^2\,\gamma^2}{(\gamma^2\,\omega^2 + \Omega^4 - 2\,\Omega^2\,\omega^2 + \omega^4)^2} + \frac{F0^{\tilde{}2}\,(\Omega^2 - \omega^2)^2}{(\gamma^2\,\omega^2 + \Omega^4 - 2\,\Omega^2\,\omega^2 + \omega^4)^2}}$$

$$\sqrt{\frac{F0^2}{m^2\omega^4 - 2\,\Omega^2 m^2\omega^2 + \Omega^4 m^2 + \gamma^2\omega^2}}$$

```
>   R(w)^2;
```

$$\frac{F0^{\tilde{}2}}{\gamma^2\,\omega^2 + \Omega^4 - 2\,\Omega^2\,\omega^2 + \omega^4}$$

```
>   f(w):=denom(%);
```

$$\Omega^4 m^2 - 2\,\Omega^2 m^2\omega^2 + m^2\omega^4 + \gamma^2\omega^2$$

```
>   eq:=diff(f(w),w)=0;
```

$$eq := \left(-4\,\Omega^2\omega + 4\,\omega^3\right)m^2 + 2\,\gamma^2\omega = 0$$

> CP:=solve(eq,w);

$$CP := 0,\ 1/2\,\frac{\sqrt{4\,\Omega^2 m^2 - 2\,\gamma^2}}{m},\ -1/2\,\frac{\sqrt{4\,\Omega^2 m^2 - 2\,\gamma^2}}{m}$$

> (CP[2])^2;

$$1/4\,\frac{4\,\Omega^2 m^2 - 2\,\gamma^2}{m^2}$$

> expand(%);

$$\Omega^2 - 1/2\,\frac{\gamma^2}{m^2}$$

> op(1,%);%/W^2;subs(W=sqrt(k/m),%);

$$-1/2\,\frac{\gamma^2}{m^2}$$

$$-1/2\,\frac{\gamma^2}{\Omega^2 m^2}$$

$$-1/2\,\frac{\gamma^2}{mk}$$

> w2[cri]:=W^2*(1+%);

$$w2_{cri} := \Omega^2\left(1 - 1/2\,\frac{\gamma^2}{mk}\right)$$

> R(w);

$$F0^{\sim}\sqrt{\frac{1}{\gamma^2\,\omega^2 + \Omega^4 - 2\,\Omega^2\,\omega^2 + \omega^4}}$$

$$F0^{\sim}\sqrt{\frac{1}{\Omega^4\,G^2\,x^2 + \Omega^4 - 2\,\Omega^4\,x^2 + x^4\,\Omega^4}}$$

> subs(F0=1,m=1,G^2=Gamma,W=1,%);

$$\sqrt{\left(x^4 + \Gamma\,x^2 - 2\,x^2 + 1\right)^{-1}}$$

$$A := (x,\Gamma) \mapsto \sqrt{\left(x^4 + \Gamma\,x^2 - 2\,x^2 + 1\right)^{-1}}$$

$$A := (x,\Gamma) \mapsto \sqrt{\left(x^4 + \Gamma\,x^2 - 2\,x^2 + 1\right)^{-1}}$$

> NewR(w):=F0/(W^2*m)*R(x,Gamma);

$$\mathrm{NewR}(\omega) := \frac{F0^{\sim}\sqrt{\dfrac{1}{\Gamma\,x^2 + 1 - 2\,x^2 + x^4}}}{\Omega^2}$$

> A:=unapply(R(x,Gamma),x,Gamma);

$$A := (x, \, \Gamma) \rightarrow \sqrt{\frac{1}{\Gamma \, x^2 + 1 - 2 \, x^2 + x^4}}$$

```
>  with(plots):
>  plot({A(x,0.2),A(x,0.5),A(x,1),A(x,1.5),A(x,
>  02)},x=0..3);
```

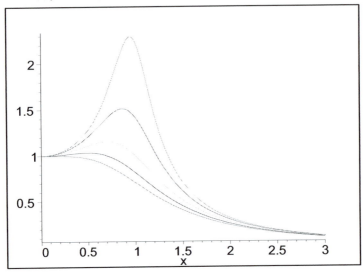

5.9.2   An Example

We recall our macros, and we use the same differential equation as before.

```
>  macro(w=omega,W=Omega,c=gamma);
>  deq:=diff(u(t),t$2)+c/m*diff(u(t),t)+W^2*u(t)=Fo/m*sin(w*t);
```

$$w, \, W, \, c$$

$$deq := (\frac{\partial^2}{\partial t^2} \, u(t)) + \frac{\gamma \, (\frac{\partial}{\partial t} \, u(t))}{m} + \Omega^2 \, u(t) = \frac{Fo \, \sin(\omega \, t)}{m}$$

Let us choose the values of the parameters in the differential equation.

$$par := \{Fo = 3 \, m, \Omega = 1, \gamma = 0.125 \, m, \omega = 2\}$$

```
>  deq1:=subs(par,deq);
```

$$deq1 := (\frac{\partial^2}{\partial t^2} \, u(t)) + .125 \, (\frac{\partial}{\partial t} \, u(t)) + u(t) = 3 \sin(2 \, t)$$

Once again we assume homogeneous initial conditions.

```
>  ics:=u(0)=0,D(u)(0)=0;
```

$$ics := u(0) = 0, \, D(u)(0) = 0$$

```
>  with(DEtools):with(plots):
```

```
Warning, the name changecoords has been redefined
```

The comand `Digits` will not effect the output of `dsolve` in MAPLE 7; however, it will set the number of digits in subsequent output.

```
> Digits:=5;
```

$$Digits := 5$$

```
> soln:=dsolve({deq1,ics},u(t));
```

$$soln := u(t) = \frac{308\, e^{-t/16}\sqrt{255}\sin\left(1/16\sqrt{255}t\right)}{2465} + \frac{12\, e^{-t/16}\cos\left(1/16\sqrt{255}t\right)}{145}$$
$$-\frac{144\sin(2t)}{145} - \frac{12\cos(2t)}{145}$$

```
> soln1:=combine(%,trig);
```

$$soln1 := u(t) = \frac{308\, e^{-t/16}\sqrt{255}\sin\left(1/16\sqrt{255}t\right)}{2465} + \frac{12\, e^{-t/16}\cos\left(1/16\sqrt{255}t\right)}{145}$$
$$-\frac{144\sin(2t)}{145} - \frac{12\cos(2t)}{145}$$

```
> u[p]:=op(3,rhs(soln1))+op(4,rhs(soln1));
```

$$u_p := -\frac{12}{145}\cos(2t) - \frac{144}{145}\sin(2t)$$

We recall the form of A and B from the previous section, and substitute the parameters into these. Note that that the output is good to 5 digits as specified.

```
> A :=
> -omega*gamma*Fo/(omega^4*m^2+(gamma^2-2*Omega^2*m^2)
> *omega^2+Omega^4*m^2);B :=
> -(-m*Omega^2+omega^2*m)*Fo/(omega^4*m^2+(gamma^2-2
> *Omega^2*m^2)*omega^2+Omega^4*m^2);
```

$$A := -\frac{\omega\,\gamma\,Fo}{\omega^4 m^2 + \left(-2\,\Omega^2 m^2 + \gamma^2\right)\omega^2 + \Omega^4 m^2}$$

$$B := -\frac{\left(-m\Omega^2 + \omega^2 m\right)Fo}{\omega^4 m^2 + \left(-2\,\Omega^2 m^2 + \gamma^2\right)\omega^2 + \Omega^4 m^2}$$

```
> a:=subs(par,A); b:=subs(par,B);
```

$$a := -.082760$$
$$b := -.99311$$

We also recall $R(\omega)$ from the previous section and substitute in the parameters.

```
> R(w):=-(-m*Omega^2+omega^2*m)*Fo/(omega^4*m^2+(gamma^2-2*
> Omega^2*m^2)*omega^2+Omega^4*m^2);
```

$$R := w \mapsto -\frac{\left(-m\Omega^2 + \omega^2 m\right)Fo}{\omega^4 m^2 + \left(-2\,\Omega^2 m^2 + \gamma^2\right)\omega^2 + \Omega^4 m^2}$$

We wish to represent the solution in the amplitude-phaseshift form

$$u(t) := R(\omega)\sin(\omega\, t + \delta)$$

```
>   R1:=subs(par,R(w));delta:=arctan(a/b);
```

$$R1 := -.99311$$
$$\delta := .083142$$

```
>   u1[p]:=-R*sin(2*t+delta);
```

$$u1_p := -R\sin(2\,t + .083142)$$

We plot the amplitude-phaseshift form of the solution against the cartesian form and as expected they lie on top of one another.

```
>   p1:=plot(rhs(soln),t=0..100):
>   p2:=plot(rhs(soln),t=0..100,style=POINT):
>   display({p1,p2});
```

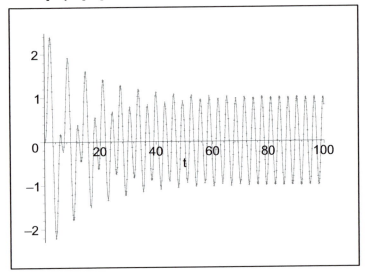

Chapter 6

Two-Point Boundary Value Problems, Catalytic Reactors and Boundary-Layer Phenomena

Boundary value problems (BVPs) are frequently encountered in mathematically describing engineering systems. Examples of BVPs include the steady-state analysis of temperature distributions, potential flow fields, diffusion, current distribution, and flow in a manifold. In this chapter, we study some of the simplest examples from catalyst analysis in chemical engineering, and boundary-layer behavior in fluid mechanics.

The process of solving a boundary value problem using Maple is practically the same as that of solving an initial-value problem, where the initial values are replaced by the boundary values. The most general boundary-value problem for a second-order linear differential equation can be formulated as: Find the function $y = y(x)$ satisfying the differential equation,

$$L[y] := \frac{d^2 y}{ax^2} + p(x)\frac{by}{ax} + q(x)y = f(x), \quad a < x < b$$

together with the boundary conditions:

$$a_0 y(a) + a_1 y'(a) = \alpha, \quad b_0 y(b) + b_1 y'(b) = \beta.$$

Here p, q, f are given functions, $a, b, a_0, a_1, b_0, b_1, \alpha$ okay and β are given constants. It is understood that the constants and the given functions are generally required to fulfill some restrictions for ensuring the existence and uniqueness of the solution to the problem.

6.1 Analytical Solutions

We begin with the simple example when the boundary conditions are of the Dirichlet type, i.e., both a_1 and b_1 are zero, but a_0, b_0 are not zero.

```
>  with(DEtools):
```

```
> deq1:=-diff(y(x),x$2)+Pi^2*y(x)=2*Pi^2*sin(Pi
> *x);
```

$$deq1 := -(\frac{\partial^2}{\partial x^2}\, \mathrm{y}(x)) + \pi^2\, \mathrm{y}(x) = 2\, \pi^2 \sin(\pi\, x)$$

```
> bcs1:=y(0)=0,y(1)=0;
```

$$bcs1 := \mathrm{y}(0) = 0,\ \mathrm{y}(1) = 0$$

```
> dsolve({deq1,bcs1},y(x));
```

$$\mathrm{y}(x) = \sin(\pi\, x)$$

Now e consider an example with a mixed boundary condition at one end. In this case we cannot obtain the solution in one step as above but must break it into two steps.

```
> deq2:=-diff(y(x),x$2)+3*diff(y(x),x)-2*y(x)=0
> ;
```

$$deq2 := -(\frac{\partial^2}{\partial x^2}\, \mathrm{y}(x)) + 3\,(\frac{\partial}{\partial x}\, \mathrm{y}(x)) - 2\, \mathrm{y}(x) = 0$$

```
> bcs2:=y(0)=1,y(1)+D(y)(1)=0;
```

$$bcs2 := \mathrm{y}(0) = 1,\ \mathrm{y}(1) + D(y)(1) = 0$$

```
> dsolve({deq2,bcs2},y(x));
```

$$\mathrm{y}(x) = -3\,\frac{e^2\, e^x}{2\, e - 3\, e^2} + \frac{2\, e\, e^{(2\, x)}}{2\, e - 3\, e^2}$$

```
> simplify(%);
```

$$\mathrm{y}(x) = \frac{(3\, e^{(2+x)} - 2\, e^{(1+2\, x)})\, e^{(-1)}}{-2 + 3\, e}$$

In case Maple does not produce the answer as in the old version then we may break the problem into two steps as follows.

```
> y2:=dsolve({deq2,y(0)=1},y(x));
```

$$y2 := \mathrm{y}(x) = (-\_C2 + 1)\, e^x + \_C2\, e^{(2\, x)}$$

```
> eq2:=diff(rhs(y2),x);
```

$$eq2 := (-\_C2 + 1)\, e^x + 2\, \_C2\, e^{(2\, x)}$$

```
> rhs(subs(x=1,y2))+subs(x=1,eq2)=0;
```

$$2\,(-\_C2 + 1)\, e + 3\, \_C2\, e^2 = 0$$

```
> _C2:=solve(%,_C2);
```

$$-C2 := -2\,\frac{e}{-2\,e + 3\,e^2}$$

> y2;

$$y(x) = \left(2\,\frac{e}{-2\,e + 3\,e^2} + 1\right)e^x - 2\,\frac{e\,e^{(2\,x)}}{-2\,e + 3\,e^2}$$

> simplify(%);

$$y(x) = \frac{\left(3\,e^{(2+x)} - 2\,e^{(1+2\,x)}\right)e^{(-1)}}{-2 + 3\,e}$$

6.2   Finite-Difference Methods

Similar to the case of IVP, in general, we are able to obtain explicit solutions only for a very small class of boundary value problems. The *finite difference method* is one of the simplest approximate methods for treating BVPs. The essential idea of the method is to approximate the derivatives of the function by the difference quotient and replace the governing differential equation by the appropriate finite difference equations. The later can then be solved by the standard methods in linear algebra. To illustrate the idea, we form a partition of the interval $[a, b]$ using the points; $a = x_0 < x_1, \cdots < \cdots x_{N+1} = b$, where $h = (b - a)/(N + 1)$ and $x_j = a + jh$ for $j = 0, 1 \cdots, N + 1$. We use the central-difference formulas to approximate the derivatives:

$$y'(x_j) = \frac{y(x_{j+1}) - y(x_{j-1})}{2h} + O(h^2),$$

and

$$y''(x_j) = \frac{y(x_{j+1}) - 2y(x_j) + y(x_{j-1})}{h^2} + O(h^2).$$

By neglecting the terms of $O(h^2)$, we replace each term $y(x_j)$ on the right hand side of the above formulas with $Y(j)$ and the resulting equations are substituted into the differential equation and the boundary conditions to obtain a system of linear algebraic equations for the unknown $Y(j), j = 0, \cdots, N + 1$.

To be more specific, using Maple we will derive the corresponding linear equations for the simple model problem:

$$-y'' + q(x)y = f(x), \quad a < x < b,$$

$$y(a) = \alpha, \quad y(b) = \beta.$$

For simplicity, we assume that $N = 4$ is fixed. The mesh size $h := \frac{b-a}{N+1} =$

$(b - a)/7$ is known. We begin with the mesh points $x[k]$ and the definitions of $q(k)$ and $f(k)$.

```
>  with(linalg):
```

Warning, new definition for norm

Warning, new definition for trace

```
>  N:=4;
```

$$N := 4$$

```
>  for k from 0 to (N+1) do
>  x[k]:=a+k*h: q(k):=q(x[k]): f(k):=f(x[k]): od:
>  for k from 1 to N do
>  eq(k):=-Y(k-1)+(2+h^2*q(k))*Y(k)-Y(k+1)=h^2*f(k);
>  od;
```

$$eq(1) := -Y(0) + (2 + h^2\, q(a + h))\, Y(1) - Y(2) = h^2\, f(a + h)$$

$$eq(2) := -Y(1) + (2 + h^2\, q(a + 2\, h))\, Y(2) - Y(3) = h^2\, f(a + 2\, h)$$

$$eq(3) := -Y(2) + (2 + h^2\, q(a + 3\, h))\, Y(3) - Y(4) = h^2\, f(a + 3\, h)$$

$$eq(4) := -Y(3) + (2 + h^2\, q(a + 4\, h))\, Y(4) - Y(5) = h^2\, f(a + 4\, h)$$

```
>  eq(1):=subs(Y(0)=alpha,eq(1));
```

$$eq(1) := -\alpha + (2 + h^2\, q(a + h))\, Y(1) - Y(2) = h^2\, f(a + h)$$

```
>  eq(N):=subs(Y(N+1)=beta,eq(N));
```

$$eq(4) := -Y(3) + (2 + h^2\, q(a + 4\, h))\, Y(4) - \beta = h^2\, f(a + 4\, h)$$

```
>  sys:=seq(eq(k),k=1..N);
```

$$\begin{aligned}
sys := {}& -\alpha + (2 + h^2\, q(a + h))\, Y(1) - Y(2) = h^2\, f(a + h), \\
& -Y(1) + (2 + h^2\, q(a + 2\, h))\, Y(2) - Y(3) = h^2\, f(a + 2\, h), \\
& -Y(2) + (2 + h^2\, q(a + 3\, h))\, Y(3) - Y(4) = h^2\, f(a + 3\, h), \\
& -Y(3) + (2 + h^2\, q(a + 4\, h))\, Y(4) - \beta = h^2\, f(a + 4\, h)
\end{aligned}$$

This system of linear equations can be solved directly and symbolicly by using the maple command solve as follows:

```
> solve(sys,seq(Y(k),k=1..N));
```
The output of the solutions is not included here. We will return to this system and discuss its matrix version at the end of the next section.

Maple Procedure

Based on the above derivation, we now develop a procedure called
```
> BVP(q,f,a,b,alpha,beta,N)
```

for our model problem so that one may obtain numerical results by varying the data parameters $q, f, a, b, alpha, beta$ and N: We consider next the differential equation

$$-\frac{d^2y(x)}{dx^2} + q(x)\,y(x) = f(x),$$

with the boundary conditions

$$y(a) = \alpha, \text{ and } y(b) = \beta,$$

where a, b, α, and β. Are given constants.

6.2.1 Finite-Difference Procedure for the Two-Point BVP:

```
> with(linalg):
> BVP:=proc(q,f,a,b,alpha,beta,N)
> local k,x,Y,h,sys,qq,ff; qq:=q: ff:=f:
> h:=(b-a)/(N+1):
> for k from 0 to (N+1) do
> x[k]:=a+k*h:
> od;
> for k from 1 to N do
> eq(k):=-Y(k-1)+(2+h^2*qq(x[k]))*Y(k)-Y(k+1)=h^2*ff(x[k])
> od:
> eq(1):=subs(Y(0)=alpha,eq(1));
> eq(N):=subs(Y(N+1)=beta,eq(N));
> sys:=evalf(seq(eq(k),k=1..N)):
> evalf(solve({sys},{seq(Y(k),k=1..N)})); end;
```

$$BVP := \mathbf{proc}(q,\ f,\ a,\ b,\ \alpha,\ \beta,\ N)$$
$$\mathbf{local}\ k,\ x,\ Y,\ h,\ sys,\ qq,\ ff;$$
$$qq := q;$$
$$ff := f;$$
$$h := (b-a)/(N+1);$$
$$\mathbf{for}\ k\ \mathbf{from}\ 0\ \mathbf{to}\ N+1\ \mathbf{do}\ x_k := a+k*h\ \mathbf{end\ do};$$
$$\mathbf{for}\ k\ \mathbf{to}\ N\ \mathbf{do}$$
$$\quad eq(k) := -Y(k-1) + (2 + h^2 * qq(x_k)) * Y(k)$$
$$\quad - Y(k+1) = h^2 * ff(x_k)$$
$$\mathbf{end\ do};$$
$$eq(1) := \mathrm{subs}(Y(0) = \alpha,\ eq(1));$$
$$eq(N) := \mathrm{subs}(Y(N+1) = \beta,\ eq(N));$$
$$sys := \mathrm{evalf}(\mathrm{seq}(eq(k),\ k = 1..N));$$
$$\mathrm{evalf}(\mathrm{solve}(\{sys\},\ \{\mathrm{seq}(Y(k),\ k = 1..N)\}))$$
$$\mathbf{end\ proc}$$

```
>   q:=x->Pi^2;
```

$$q := x \to \pi^2$$

```
>   f:=x->2*Pi^2*sin(Pi*x);
```

$$f := x \to 2\,\pi^2 \sin(\pi\,x)$$

```
>   Y:=BVP(q,f,0,1,0,0,6);
```

$Y := \{Y(6) = .4375311633,\ Y(3) = .9831236000,\ Y(5) = .7884039138,$
$Y(4) = .9831236000,\ Y(1) = .4375311633,\ Y(2) = .7884039138\}$

```
>   sin(Pi/7);
```

$$\sin(\tfrac{1}{7}\pi)$$

```
>   evalf(%);
```

$$.4338837393$$

Let us check if doubling the number of mesh points increases the accuracy.

```
>   Y:=BVP(q,f,0,1,0,0,13);
```

$Y := \{Y(7) = 1.002099003,\ Y(10) = .7834725494,\ Y(9) = .9028600046,$
$Y(6) = .9769742891,\ Y(8) = .9769742891,\ Y(11) = .6247985091,$
$Y(13) = .2229880062,\ Y(12) = .4347944626,\ Y(3) = .6247985091,$
$Y(2) = .4347944626,\ Y(1) = .2229880062,\ Y(4) = .7834725494,$
$Y(5) = .9028600046\}$

We remark that in general the system of linear equations from the finite-difference methods are usualy solved by matrix methods. For this purpose, we show how one may also generate the coefficient matrix and the augmented matrix by maple commands such as `genmatrix(eqns, vars)` and `genmatrix(eqns,vars,flag)`: Here the parameters of the commands are:

eqns:–set or list of equations

vars:–set or list of variables

flag:–(option) a flag which can be any values

> `sys:=seq(eq(k),k=1..N);`

$$sys := -\alpha + (2 + h^2 q(a + h)) Y(1) - Y(2) = h^2 f(a + h),$$
$$-Y(1) + (2 + h^2 q(a + 2h)) Y(2) - Y(3) = h^2 f(a + 2h),$$
$$-Y(2) + (2 + h^2 q(a + 3h)) Y(3) - Y(4) = h^2 f(a + 3h),$$
$$-Y(3) + (2 + h^2 q(a + 4h)) Y(4) - \beta = h^2 f(a + 4h)$$

> `A:=genmatrix([sys],[seq(Y(k),k=1..N)]);`

$$A := \begin{bmatrix} 2 + h^2 q(a + h) & -1 & 0 & 0 \\ -1 & 2 + h^2 q(a + 2h) & -1 & 0 \\ 0 & -1 & 2 + h^2 q(a + 3h) & -1 \\ 0 & 0 & -1 & 2 + h^2 q(a + 4h) \end{bmatrix}$$

> `Ab:=genmatrix([sys],[seq(Y(k),k=1..N)],flag);`

$$Ab := \begin{bmatrix} 2 + h^2 q(a + h), -1, 0, 0, \alpha + h^2 f(a + h) \\ -1, 2 + h^2 q(a + 2h), -1, 0, h^2 f(a + 2h) \\ 0, -1, 2 + h^2 q(a + 3h), -1, h^2 f(a + 3h) \\ 0, 0, -1, 2 + h^2 q(a + 4h), \beta + h^2 f(a + 4h) \end{bmatrix}$$

One may use the standard methods of linear algebra to handle augmented matrices. Maple commands such as `gausselim(Ab)` and `gaussjord(Ab)` are particularly designed for this purpose. By virtue of these commands, it is now rather straightforward how to modify the BVP-procedure with a matrix method to solve the corresponding system of linear equations.

6.3 Computer Lab

Boundary-Layer Phenomenon: Consider the boundary value problem $(BVP)_\varepsilon$:

$$\varepsilon y''(x) + y'(x) + y(x) = 0, \quad 0 < x < 1,$$

$$y(0) = 0, \quad y(1) = 1,$$

where $0 < \varepsilon << 1$ is a small parameter.

- Find the exact solution y_ε of the $(BVP)_\varepsilon$.

- Solve the reduced problem $(BVP)_0$ defined by

$$y_0' + y_0 = 0, \quad 0 < x < 1, \quad y_0(1) = 1.$$

- Sketch in the same graphs of y_0 and y_ε for $\varepsilon = 0.1, 0.05, 0.01$. You may notice that the reduced solution y_0 is a good approximation of y_ε except in a narrow region:

$$0 \leq x \leq x_\varepsilon.$$

This narrow region is an example of the boundary-layer——a region of sudden transition. Boundary layer phenomena occur in many problems of fluid mechanics and in other parts of mathematical physics. A great deal of work has been done in the late 60s unraveling some of the many perplexing features of equations of the foregoing type, but much more remains to be done. We will return to this topic again in a later time for a little more detail.

Catalytic Reactor: A common practice in heterogeneous catalysis is the use of the catalyst in the form of particles, ranging in size from a power to good-size pills, often artificially made. Diffusion and first-order reaction inside an isothermal spherical catalyst particle can be modeled by the boundary-value problem in dimensionless form for the concentration c :

$$c''(r) + \frac{2}{r}c'(r) = \Phi^2 c(r), \quad 0 < r < 1,$$

$$c'(0) = 0, c(1) = 1,$$

where Φ^2 is a given constant, the so-called *Thiele modules*, and is proportional to the reaction rate constant. The solution of this linear, second-order differential equation with one non-constant coefficient can be expressed in terms of Bessel functions. We now consider its numerical approximations.

- Derive the finite-difference equations for this boundary-value problem by assuming that at $r = 0$, c satisfies the limiting equation:

$$-c''(r) + \frac{\Phi^2}{3}c(r) = 0, \quad r = 0.$$

- Modify the Maple procedure given previously so that it can be apply to the present model problem.

- Compute approximate solutions for $\Phi = 0.1, 1, 5$ and 10 by taking $N = 9$. Plot the exact solution

$$c(r) = \frac{1}{r} \frac{sinh(\Phi r)}{sinh(\Phi)}$$

and the approximate solutions on the same graph by varying Φ. Can you make any comment on the influence of Φ with respect to the accuracy of the numerical solutions ?

6.4 Supplementary Maple Programs

6.4.1 An Exact Asymptotic Expansion

```
> with(DEtools):
> deq:=epsilon*diff(u(t),t$2)+diff(u(t),t)+u(t)=0;
```

$$deq := \varepsilon \left(\tfrac{\partial^2}{\partial t^2} \, u(t) \right) + \left(\tfrac{\partial}{\partial t} \, u(t) \right) + u(t) = 0$$

```
> bcs:=u(0)=0,u(1)=1;
```

$$bcs := u(0) = 0, \ u(1) = 1$$

```
> macro(a=epsilon);
```

$$a$$

```
> eq:=a*r^2+r+1=0;
```

$$eq := \varepsilon \, r^2 + r + 1 = 0$$

```
> solve(eq,r);
```

$$\frac{1}{2} \frac{-1 + \sqrt{1 - 4\varepsilon}}{\varepsilon}, \ \frac{1}{2} \frac{-1 - \sqrt{1 - 4\varepsilon}}{\varepsilon}$$

```
> r1:=unapply(%[1],a); r2:=unapply(%%[2],a);
```

$$r1 := \varepsilon \to \frac{1}{2} \frac{-1 + \sqrt{1 - 4\varepsilon}}{\varepsilon}$$

$$r2 := \varepsilon \to \frac{1}{2} \frac{-1 - \sqrt{1 - 4\varepsilon}}{\varepsilon}$$

```
> series(r1(a),a=0,3);
```

$$-1 - \varepsilon + O(\varepsilon^2)$$

```
> ar1:=convert(%,polynom);
```

$$ar1 := -1 - \varepsilon$$

```
> series(r2(a),a=0,3);
```

$$-\varepsilon^{-1} + 1 + \varepsilon + O(\varepsilon^2)$$

```
> ar2:=convert(%,polynom);
```

$$ar2 := -\frac{1}{\varepsilon} + 1 + \varepsilon$$

```
>  soln:=c1*exp(ar1*t)+c2*exp(ar2*t);
```

$$soln := c1\, e^{((-1-\varepsilon)\,t)} + c2\, e^{((-\frac{1}{\varepsilon}+1+\varepsilon)\,t)}$$

```
>  bc1:=subs(t=0,soln)=0;
```

$$bc1 := c1\, e^0 + c2\, e^0 = 0$$

```
>  bc2:=subs(t=1,soln)=1;
```

$$bc2 := c1\, e^{(-1-\varepsilon)} + c2\, e^{(-\frac{1}{\varepsilon}+1+\varepsilon)} = 1$$

```
>  sols:=solve({bc1,bc2},{c1,c2});
```

$$sols := \left\{ c2 = -\frac{1}{e^{(-1-\varepsilon)} - e^{(\frac{-1+\varepsilon+\varepsilon^2}{\varepsilon})}},\; c1 = \frac{1}{e^{(-1-\varepsilon)} - e^{(\frac{-1+\varepsilon+\varepsilon^2}{\varepsilon})}} \right\}$$

```
>  assign(sols);
>  AsyS:=normal(soln);
```

$$AsyS := \frac{e^{(-(1+\varepsilon)\,t)} - e^{(\frac{(-1+\varepsilon+\varepsilon^2)\,t}{\varepsilon})}}{e^{(-1-\varepsilon)} - e^{(\frac{-1+\varepsilon+\varepsilon^2}{\varepsilon})}}$$

```
>  U[a]:=exp(1-t)-exp(1+t)*exp(-t/a);
```

$$U_\varepsilon := e^{(1-t)} - e^{(1+t)}\, e^{(-\frac{t}{\varepsilon})}$$

```
>  with(plots):
```

Warning, the name changecoords has been redefined

```
>  p1:=plot(subs(a=0.01,AsyS),t=0..1,style=POINT,symbol=CIRCLE):
>  p2:=plot(subs(a=0.01,U[a]),t=0..1,style=POINT,symbol=CROSS):
>  ExS:=normal(dsolve({deq,bcs},u(t)));
```

$$ExS := u(t) = \frac{e^{(1/2\,\frac{(-1+\%1)\,t}{\varepsilon})} - e^{(-1/2\,\frac{(1+\%1)\,t}{\varepsilon})}}{e^{(1/2\,\frac{-1+\%1}{\varepsilon})} - e^{(-1/2\,\frac{1+\%1}{\varepsilon})}}$$

$$\%1 := \sqrt{1 - 4\varepsilon}$$

```
>  p3:=plot(rhs(subs(a=0.01,ExS)),t=0..1):
>  display({p1,p3});
```

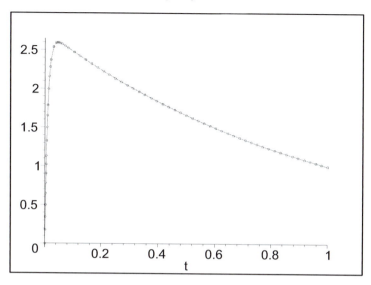

```
>   display({p2,p3});
```

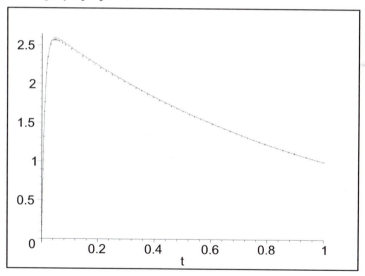

We remark that for ε sufficiently small, it can be shown that U_ε is indeed a good approximation of the exact solution u in the sense that

$$u(t) = U_\varepsilon(t) + O(\varepsilon)$$

uniformly for all $t \in [0, 1]$. In terms of the terminology of *singular perturbation theory*, the second term in U_ε, namely,

$$-exp(1+t)exp(-\frac{t}{\varepsilon})$$

is called the boundary-layer correction (or the zero-th order term in the *inner expansion*) which is only significant in the neighbourhood of boundary-layer (i.e., of t = 0 in the present example).

In the above figures, the symbols ○ and × denote the graphs of *AsyS* and U_ε, respectively and the solid line is the exact solution *u*.

Chapter 7

Eigenvalue Problems

An eigenvalue problem for a linear operator L is the problem of finding those real or complex values of the parameter λ, called *eigenvalues*, for which the equation $L[y] = \lambda y$ has a *nontrivial solution* ($y \neq 0$) in the domain of the operator L. The corresponding solutions are called *eigenfunctions* (or *eigensolutions*). The topics of eigenvalues is important and arises in many places in mathematics and engineering. The word *eigenvalues* itself is a hybrid German and English word meaning *special* values. The purposes of this chapter is to discuss some general properties concerning eignvalue problems via simple model problems and introduce elementary approximate methods for constructing their solutions.

7.1 Sturm-Liouville Problems

In the elementary text, when the operator L is a differential operator, the eigenvalue problem is usually formulated in the form of a homogeneous boundary-value problem. In this context the term *homogeneous problem* means that if we can find a solution of the problem then any multiple of that solution is also a solution.

As a simple example, we begin with the eigenvalue problem (EVP):

$$y'' + \lambda y = 0, \quad 0 < x < 1,$$

$$y(0) = 0, \quad y(1) = 0.$$

The special values of λ,

$$\lambda_n := (n\pi)^2, \, n = 1, 2, \cdots,$$

are the eigenvalues and the corresponding eigenfunctions are

$$y_n = A \, sin(\sqrt{\lambda_n}x) \quad \text{for any constant } A \neq 0.$$

In term of the differential operator, we see that

$$L[y](x) := -y''(x),$$

and the domain of L is the set of functions with continuous second derivatives on the interval $[0, 1]$ that satisfy the boundary conditions $y(0) = 0$, $y(1) = 0$.

This simple EVP belongs to a general class of homogeneous boundary-value problems called the *Sturm-Liouville Problems* (SLP) consist of differential equations of the form:

$$L[y] := -(p(x)y')' + q(x)y = \lambda r(x)y, \quad 0 < x < 1$$

together with boundary conditions,

$$a_1 y(0) + a_2 y'(0) = 0, \quad b_1 y(1) + b_2 y'(1) = 0.$$

Here p, p', and r are given continuous functions on the interval $0 \leq x \leq 1$, and $p(x) > 0$ and $r(x) > 0$ for all $x \in [0, 1]$.

Theorem 2. *The eigenvalues and eigenfunctions of the* SLP *have the following properties:*

(a) *The eigenvalues are all real and nonnegative.*

(b) *The eigenvalues can be arranged to form a strictly increasing infinite sequence; that is, $0 \leq \lambda_1 < \lambda_2 < \lambda_3 < \cdots$. Furthermore, $\lambda_n \to \infty$ as $n \to \infty$.*

(c) *The eigenvalues are all simple; that is, to each eigenvalue, there corresponds only one linearly independent eigenfunction.*

(d) *The set of linearly independent eigenfunctions $\{\varphi_1(x), \varphi_2(x), \cdots\}$ satisfies the orthogonal relation:*

$$\int_0^1 r(x)\varphi_m(x)\varphi_n(x)dx = 0 \quad \text{for} \quad m \neq n.$$

As one can see, all these properties can be verified directly via the simple model problem (EVP) Chapter 7.

7.2   Numerical Approximations

We now consider some aspects of the numerical solutions of the SLPs. To illustrate the idea, we return to the simple model problem (EVP). It will be convenient to let $h := \frac{1}{N+1}$, so the interval $[0, 1]$ is subdivided into $N + 1$ subintervals of length h. In the usual way, the differential equation is replaced by the central finite-difference equations. This can be easily achieved by using Maple language.

```
>  with(linalg): with(DEtools): with(plots):

Warning, the name adjoint has been redefined
```

Warning, the name adjoint has been redefined
```
>   deq:=-diff(y(x),x$2)=lambda*y(x);
```

$$deq := -(\frac{\partial^2}{\partial x^2} \, y(x)) = \lambda^\sim y(x)$$

```
>   bcs:=y(0)=0,y(1)=0;
```

$$bcs := y(0) = 0, \; y(1) = 0$$

```
>   N:=3;
```

$$N := 3$$

```
>   for k from 0 to N+1 do
>   #h:=1/(N+1);
>   x[k]:=k*h;
>   #Y(k):=y(x[k]);
>   od:
>   for k from 1 to N do
>   eq(k):=-(Y(k-1)-2*Y(k)+Y(k+1))/h^2=lambda*Y(k);
>   od;
```

$$eq(1) := -\frac{Y(0) - 2\,Y(1) + Y(2)}{h^2} = \lambda^\sim Y(1)$$

$$eq(2) := -\frac{Y(1) - 2\,Y(2) + Y(3)}{h^2} = \lambda^\sim Y(2)$$

$$eq(3) := -\frac{Y(2) - 2\,Y(3) + Y(4)}{h^2} = \lambda^\sim Y(3)$$

```
>   eq(1):=subs(Y(0)=0,eq(1));
```

$$eq(1) := -\frac{-2\,Y(1) + Y(2)}{h^2} = \lambda^\sim Y(1)$$

```
>   eq(N):=subs(Y(N+1)=0,eq(N));
```

$$eq(3) := -\frac{Y(2) - 2\,Y(3)}{h^2} = \lambda^\sim Y(3)$$

```
>   sys:=seq(eq(k),k=1..N);
```

$$sys := -\frac{-2\,Y(1) + Y(2)}{h^2} = \lambda^\sim Y(1), \; -\frac{Y(1) - 2\,Y(2) + Y(3)}{h^2} = \lambda^\sim Y(2),$$
$$-\frac{Y(2) - 2\,Y(3)}{h^2} = \lambda^\sim Y(3)$$

```
>   vars:=seq(Y(k),k=1..N);
```

$$vars := Y(1), \; Y(2), \; Y(3)$$

```
>   A:=GenerateMatrix([sys],[vars]);
```

$$A := \begin{bmatrix} 2/h^2 - \lambda & -1/h^2 & 0 \\ -1/h^2 & 2/h^2 - \lambda & -1/h^2 \\ 0 & -1/h^2 & 2/h^2 - \lambda \end{bmatrix},$$

$$\begin{bmatrix} 0 \\ 0 \\ 0 \end{bmatrix}$$

> `Determinant(A[1]);`

$$-\frac{h^6 \lambda^3 - 6 h^4 \lambda^2 + 10 \lambda h^2 - 4}{h^6}$$

> `subs(h=1/(N+1),%);`

$$-\lambda^3 + 96 \lambda^2 - 2560 \lambda + 16384$$

> `p:=unapply(%,lambda);`

$$p := \lambda \mapsto -\lambda^3 + 96 \lambda^2 - 2560 \lambda + 16384$$

> `Lambda:=evalf(solve(p(lambda)=0,lambda));`

$$\Lambda := 32., 54.62741699, 9.37258301, 54.62741699,$$

> `lambda[ex]:=evalf(seq((n*Pi)^2,n=1..3));`

$$\lambda_{ex} := 9.869604404, 39.47841762, 88.82643964$$

> `plot(p(lambda),lambda=0..9*Pi^2);`

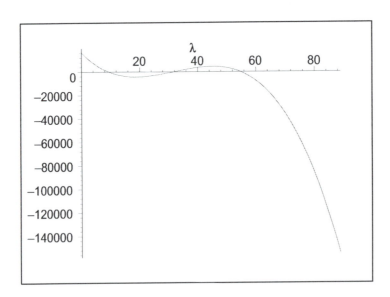

> `sys1:=seq(lhs(eq(k)),k=1..N);`

$$sys1 := -\frac{-2\,Y(1) + Y(2)}{h^2}, \; -\frac{Y(1) - 2\,Y(2) + Y(3)}{h^2}, \; -\frac{Y(2) - 2\,Y(3)}{h^2}$$

```
>   A:=GenerateMatrix([sys1],[seq(Y(k),k=1..N)]);
```

$$A := \begin{bmatrix} 2/h^2 & -1/h^2 & 0 \\ -1/h^2 & 2/h^2 & -1/h^2 \\ 0 & -1/h^2 & 2/h^2 \end{bmatrix}, \begin{bmatrix} 0 \\ 0 \\ 0 \end{bmatrix}$$

```
>   AA:=subs(h=1/(N+1),op(A));
```

$$AA := \begin{bmatrix} 32 & -16 & 0 \\ -16 & 32 & -16 \\ 0 & -16 & 32 \end{bmatrix}$$

```
>   Eigenvalues(AA);
```

$$\begin{bmatrix} 32 \\ 32 - 16\sqrt{2} \\ 32 + 16\sqrt{2} \end{bmatrix}$$

```
>   Lambda:= {%};
```

$$\Lambda := \begin{bmatrix} 32 \\ 32 - 16\sqrt{2} \\ 32 + 16\sqrt{2} \end{bmatrix}$$

```
>   ES1:=NullSpace(CharacteristicMatrix(AA,Lambda[3]));
```

$$ES1 := \left\{ \begin{bmatrix} 1 \\ -\sqrt{2} \\ 1 \end{bmatrix} \right\}$$

```
>   V1:=op(%);
```

$$V1 := \begin{bmatrix} 1 \\ -\sqrt{2} \\ 1 \end{bmatrix}$$

```
>   ES2:=NullSpace(CharacteristicMatrix(AA,Lambda[1]));
```

$$ES2 := \left\{ \begin{bmatrix} -1 \\ 0 \\ 1 \end{bmatrix} \right\}$$

```
>   V2:=op(%);
```

$$V2 := \begin{bmatrix} -1 \\ 0 \\ 1 \end{bmatrix}$$

```
>   ES3:=NullSpace(CharacteristicMatrix(AA,Lambda[2]));
```

$$ES3 := \left\{ \begin{bmatrix} 1 \\ \sqrt{2} \\ 1 \end{bmatrix} \right\}$$

```
>  V3:=op(%);
```

$$V3 := \begin{bmatrix} 1 \\ \sqrt{2} \\ 1 \end{bmatrix}$$

```
>  Y1:=[seq([k/(N+1),V1[k]], k=1..N)]:
>  p1:=plot(sqrt(2)*sin(Pi*x),x=0..1):
>  ap1:=plot(Y1,style=POINT,symbol=CIRCLE):
>  display({p1,ap1});
```

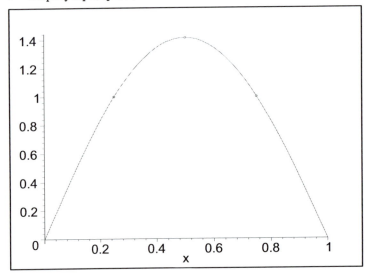

```
>  Y2:=[seq([k/(N+1),V2[k]], k=1..N)]:
>  p2:=plot(-sin(2*Pi*x),x=0..1):
>  ap2:=plot(Y2,style=POINT,symbol=CIRCLE):
>  display({p2,ap2});
>  Y3:=[seq([k/(N+1),V3[k]], k=1..N)]:
```

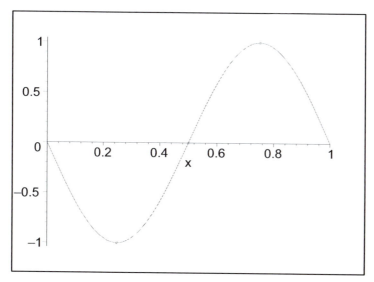

```
>   p3:=plot(sqrt(2)*sin(3*Pi*x),x=0..1):
>   ap3:=plot(Y3,style=POINT,symbol=CIRCLE):
>   display({p3,ap3});
```

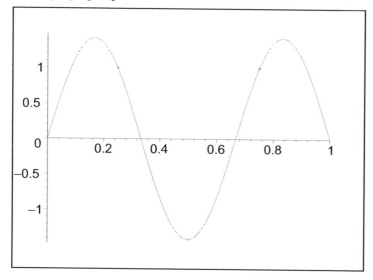

If we look at the numbers, we see that the approximation is fairly good for λ_1, less good for λ_2 and least good for λ_3. This is typical of the situation. If we let $\lambda_1 < \lambda_2 < \lambda_3 < \cdots$ be the eigenvalues of the EVP as in the theorem above, then the difference equations have exactly N eigenvalues, and these approximate the N smallest $\lambda's$, $\lambda_1, \lambda_2, \cdots, \lambda_N$. The accuracy of the approximate Λ_n to λ_n decreases as n increases.

7.3   The Newton-Raphson Method

To obtain approximate eigenvalues, one needs in general to find the roots of the characteristic equation of the coefficient matrix A; that is,

$$p(\Lambda) := det(A - \Lambda I) = 0.$$

In fact, one of the most frequently occurring problems in scientific work is to find the roots of equation of the form:

$$f(x) = 0.$$

The Newton-Raphson method is one of the classical iteration methods which is particularly desirable for this purpose. The algorithm begins with an initial guess x_0 for the root. Then it will be improved successively by a sequence $\{x_n\}$ defined by

$$x_n := x_{n-1} - \frac{f(x_{n-1})}{f'(x_{n-1})}, \quad n = 1, 2, \cdots,$$

provided $f'(x_{n-1}) \neq 0$. It is expected that the sequence $\{x_n\}$ will eventually converge to a root of the equation $f(x) = 0$, if the initial guess x_0 is sufficiently close to the root. Indeed, given below is a set of sufficient conditions for the convergence of x_n.

Theorem 3. *Suppose that the equation $f(x) = 0$ has a root at x^*, with $f'(x^*) \neq 0$ and $f''(x^*)$ continuous at x^*. Then there exists a $\delta > 0$ such that if $|x^* - x_0| \leq \delta$ then the sequence given by the Newton-Raphson algorithm converges to x^*.*

In the following we give our Maple version of the Newton-Raphson iteration scheme as a procedure, NEWTON(f,startvalue,tolerance).

```
> with(linalg):
```

Warning, the protected names norm and trace have been redefined and unprotected

```
> NEWTON:=proc(f,startvalue,tolerance)
> description "This is the Newton=Raphson method.";
> local x0,x1; g:=f;
> x0:=startvalue;
> while abs(f(x0)) > tolerance do
```

```
>  x1:=evalf(x0-g(x0)/D(g)(x0));
>  x0:=x1:
>  print('approx x:=',x0) od;
>  RETURN(x0);
>  end;
```

Warning, 'g' is implicitly declared local to procedure 'NEWTON'

$NEWTON :=$ **proc**(f, *startvalue*, *tolerance*)
local $x0$, $x1$, g;
description "This is the Newton=Raphson method.";
   $g := f$;
   $x0 := startvalue$;
   while $tolerance <$ abs($f(x0)$) **do**
      $x1 := $ evalf$(x0 - g(x0)/\mathrm{D}(g)(x0))$; $x0 := x1$; print('*approx x := '*, $x0$)
   end do;
   $RETURN(x0)$
end proc
```
>  p:=x-> -x^3 +96*x^2 -2560*x + 16384;
```
$$p := x \rightarrow -x^3 + 96\,x^2 - 2560\,x + 16384$$
```
>  NEWTON(p,8,10^(-6));
```
$$approx\ x :=, 9.263157895$$
$$approx\ x :=, 9.371798097$$
$$approx\ x :=, 9.372582960$$
$$approx\ x :=, 9.372583000$$
$$approx\ x :=, 9.372583002$$
$$approx\ x :=, 9.372582997$$
$$approx\ x :=, 9.372583005$$
$$9.372583005$$
```
>  save NEWTON,"NEWTON.m":
```

We remark that in order to make a good choice of the initial guess, it is wise to plot the function first and then obtain, as the inital guess, an approximate root from the graph of the function.

7.4 Computer Lab

Consider the following eigenvalue problem:

$$-y'' = \lambda y, \quad 0 < x < 1,$$

$$y(0) + y'(0) = 0, \quad y(1) = 0.$$

- Show that the eigenvalues are solutions of the transcendental equation

$$tan(\sqrt{\lambda}) = \sqrt{\lambda},$$

by following the steps outlined below:
(a) Express the solution of the differential equation in the form $y = c_1 cos(\sqrt{\lambda}x) + c_2 sin(\sqrt{\lambda}x)$. (b) Use the boundary conditions to obtain a system of two linear equations for the unknown coefficients. (c) Set the determinant of the coefficient matrix equal to zero.

- Plot the curves

$$y_1 = x \quad \text{and} \quad y_2 = tan(x)$$

to locate their intersects approximately from the graph. As you will see, there are infinite number of eigenvalues. They are approaching the square of the odd multiples of $\pi/2$; that is, $\lambda_n \simeq (2n+1)^2\pi^2/4$. Find the first three eigenvalues from the graph.

- Find the first eigenvalue λ_1 by the Newton-Raphson method with a tolerance: $\varepsilon = 10^{-4}$; that is,

$$|f(\lambda_1)| \leq 10^{-4}$$

in the iteration, where $f(\lambda) = tan(\sqrt{\lambda}) - \sqrt{\lambda}$. Hint: Choose an initial guess properly from the above graph.

- Modify the finite-difference scheme from (EVP) so that it can be adapted for the present problem. Find the first three approximate eigenvalues by the scheme with $N = 3$. By comparing them to the graph solutions of the transcendental equation, what conclusions can you make ?

7.5 Supplementary Mapple Programs

7.5.1 An Eigenvalue Equation

```
> restart:
```

```
>  with(linalg): with(plots): with(DEtools):
```

Warning, the protected names norm and trace have been redefined
and unprotected

Warning, the name changecoords has been redefined

Warning, the name adjoint has been redefined

```
>  macro(c=lambda):
>  assume(c>0);
>  deq:=diff(y(x),x$2)+c*y(x)=0;
```

$$deq := \frac{\mathrm{d}^2}{\mathrm{d}x^2}y(x) + \lambda\, y(x) = 0$$

```
>  bcs:=y(0)+D(y)(1)=0,y(1)=0;
```

$$bcs := y(0) + \mathrm{D}(y)(1) = 0,\ y(1) = 0$$

```
>  soln:=dsolve(deq,y(x));
```

$$soln := y(x) = \_C1\,\sin\left(\sqrt{\lambda}x\right) + \_C2\,\cos\left(\sqrt{\lambda}x\right)$$

```
>  psoln:=diff(soln,x);
```

$$psoln := \frac{\mathrm{d}}{\mathrm{d}x}y(x) = \_C1\,\sqrt{\lambda}\cos\left(\sqrt{\lambda}x\right) - \_C2\,\sqrt{\lambda}\sin\left(\sqrt{\lambda}x\right)$$

```
>  eq1:=subs(x=1,rhs(soln))=0;
```

$$eq1 := \_C1\,\sin\left(\sqrt{\lambda}\right) + \_C2\,\cos\left(\sqrt{\lambda}\right) = 0$$

```
>  eq2:=simplify(subs(x=0,rhs(psoln))+subs(x=0,rhs(soln)))=0;
```

$$eq2 := \_C1\,\sqrt{\lambda} + \_C2 = 0$$

```
>  A:= GenerateMatrix([eq1,eq2],[_C1,_C2]);
```

$$A := \left[\begin{array}{cc} sin(\sqrt{(\lambda)}) & cos(\sqrt{(\lambda)}) \\ \lambda & 1 \end{array}\right], \left[\begin{array}{c} 0 \\ 0 \end{array}\right]$$

$$\sin(\sqrt{\lambda^{\sim}}) - \cos(\sqrt{\lambda^{\sim}})\sqrt{\lambda^{\sim}} = 0$$

```
>  %/cos(sqrt(c));
```

$$\frac{\sin\left(\sqrt{\lambda}\right) - \cos\left(\sqrt{\lambda}\right)\sqrt{\lambda}}{\cos\left(\sqrt{\lambda}\right)} = 0$$

```
>  expand(%);
```

$$\frac{\sin\left(\sqrt{\lambda}\right)}{\cos\left(\sqrt{\lambda}\right)} - \sqrt{\lambda} = 0$$

```
>  subs(op(1,lhs(%))=tan(sqrt(c)),%);
```

$$\tan\left(\sqrt{\lambda}\right) - \sqrt{\lambda} = 0$$

```
>  plot({x,tan(x)},x=0..4*Pi,y=-Pi..4*Pi);
```

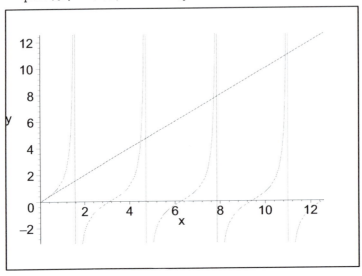

```
>  x1:=4.555: x2:=7.76: x3:=10.93:
>  f:=x->tan(sqrt(x))-sqrt(x);
```

$$f := x \mapsto \tan\left(\sqrt{x}\right) - \sqrt{x}$$

```
>  read "NEWTON.m":
>  epsilon:=10^(-4):
>  NEWTON(f,x1^2,epsilon);
```

$approx\ x :=,\ 20.34738100$

$approx\ x :=,\ 20.20323994$

$approx\ x :=,\ 20.19080873$

$approx\ x :=,\ 20.19072856$

20.19072856

```
>  NEWTON(f,x2^2,epsilon);
```

$approx\ x :=,\ 59.82474031$

$approx\ x :=,\ 59.69013272$

$approx\ x :=,\ 59.67957277$

$approx\ x :=,\ 59.67951594$

59.67951594

```
>  NEWTON(f,x3^2,epsilon);
```

$approx\ x :=,\ 119.0597354$

$approx\ x :=,\ 118.9126904$

$approx\ x :=,\ 118.8999517$

$approx\ x :=,\ 118.8998693$

$$118.8998693$$

```
>  fsolve(f(x)=0,x, 60..120);
```

$$118.8998692$$

We remark that the roots of the eigenvalue equations can also be obtained directly by using the command `fsolve`.

Chapter 8

Power Series Methods for Solving Differential Equations

In this chapter we show how to use Maple to obtain series solutions to differential equations with analytic coefficients. For the first example, we consider a second-order differential equation with **analytic** coefficients, and obtain an expansion about a **regular** point. The coefficient $\sin x$ is analytic at all points; hence, $x = 0$ is a regular point.

```
>   ode1:=
>   diff(y(x),x$2)+sin(x)*y(x);
```

$$ode1 := (\frac{\partial^2}{\partial x^2} \, y(x)) + \sin(x) \, y(x)$$

We substitute in the truncated series

$$y(x) \;=\; \sum_{k=0}^{N} a_k x^k$$

and then simplify.

```
>   p:=simplify(subs(y(x)
>   =sum(a(k)*x^k,k=0..8),ode1));
```

$$
\begin{aligned}
p := \; & 2\,a(2) + 6\,a(3)\,x + 12\,a(4)\,x^2 + 20\,a(5)\,x^3 + 30\,a(6)\,x^4 + 42\,a(7)\,x^5 + 56\,a(8)\,x^6 \\
& + \sin(x)\,a(0) + \sin(x)\,a(1)\,x + \sin(x)\,a(2)\,x^2 + \sin(x)\,a(3)\,x^3 + \sin(x)\,a(4)\,x^4 \\
& + \sin(x)\,a(5)\,x^5 + \sin(x)\,a(6)\,x^6 + \sin(x)\,a(7)\,x^7 + \sin(x)\,a(8)\,x^8
\end{aligned}
$$

In order to collect terms effectively and truncate terms we take the Taylor series of the above output.

```
>   p:=taylor(p,x,8);
```

$$p := 2\,a(2) + (6\,a(3) + a(0))\,x + (12\,a(4) + a(1))\,x^2 + (20\,a(5) - \frac{1}{6}\,a(0) + a(2))\,x^3 +$$

$$(-\frac{1}{6}\,a(1) + 30\,a(6) + a(3))\,x^4 + (a(4) + 42\,a(7) + \frac{1}{120}\,a(0) - \frac{1}{6}\,a(2))\,x^5 +$$

$$(56\,a(8) + \frac{1}{120}\,a(1) + a(5) - \frac{1}{6}\,a(3))\,x^6 + (-\frac{1}{6}\,a(4) +$$

$$a(6) - \frac{1}{5040}\,a(0) + \frac{1}{120}\,a(2))\,x^7 + O(x^8)$$

We shall seek two solutions one with the initial conditions

$$y(0) = 1 \text{ and } \frac{dy}{dx}(0) = 0$$

and the other with the initial conditions

$$y(0) = 0 \text{ and } \frac{dy}{dx}(0) = 1.$$

It is well-known that these two solutions will be linear independent as their Wronskian is not zero at $x = 0$. To solve for the **Taylor coefficients** for the first series solution, we set $a_0 = 1$ and $a_1 = 0$. Having done this we collect the coefficients of the various powers of x, by using the command `coeff`. To collect these as a list this is done using a `do` loop.

> `p1:=subs({a(0)=1,a(1)=0},p);`

$$p1 := 2\,a\,(2) + (6\,a\,(3) + 1)\,x + 12\,a\,(4)\,x^2 + (20\,a\,(5) + a\,(2) - 1/6)\,x^3 +$$

$$(30\,a\,(6) + a\,(3))\,x^4 + \left(42\,a\,(7) + a\,(4) - 1/6\,a\,(2) + \frac{1}{120}\right)x^5$$

$$+ (56\,a\,(8) + a\,(5) - 1/6\,a\,(3))\,x^6 + \left(a\,(6) - 1/6\,a\,(4) + \frac{a\,(2)}{120} - \frac{1}{5040}\right)$$

$$x^7 + O\left(x^8\right)$$

> `for n from 0 to 8 do eq(n):=coeff(p1,x,n) od;`

$$\text{eq}(0) := 2\,a(2)$$
$$\text{eq}(1) := 6\,a(3) + 1$$
$$\text{eq}(2) := 12\,a(4)$$
$$\text{eq}(3) := 20\,a(5) - \frac{1}{6} + a(2)$$
$$\text{eq}(4) := 30\,a(6) + a(3)$$
$$eq\,(5) := 42\,a\,(7) + a\,(4) - 1/6\,a\,(2) + \frac{1}{120}$$
$$eq\,(6) := 56\,a\,(8) + a\,(5) - 1/6\,a\,(3)$$

$$eq\,(7) := a\,(6) - 1/6\,a\,(4) + \frac{a\,(2)}{120} - \frac{1}{5040}$$

$$eq(8) := O(1)$$

We then solve the first six equations for the coefficients a_k, $k = 2\ldots 7$.

```
>  l:=solve({eq(0),eq(1),eq(2),eq(3),eq(4),eq(5)},
>  {a(2),a(3),a(4),a(5),a(6),a(7)});
```

$$l := \left\{ a\,(2) = 0, a\,(3) = -1/6, a\,(4) = 0, a\,(5) = \frac{1}{120}, a\,(6) = \frac{1}{180}, a\,(7) = -\frac{1}{5040} \right\}$$

These coefficients are substituted into the power series representation.

```
>  y1:=subs({a(4) = 0, a(5) = 1/120, a(6) =
>  1/180, a(7) = -1/5040, a(2)
>  = 0, a(3) = -1/6},1+sum(a(k)*x^k,k=2..7));
```

$$y1 := 1 - \frac{1}{6}x^3 + \frac{1}{120}x^5 + \frac{1}{180}x^6 - \frac{1}{5040}x^7$$

Next we plot the Taylor solution and compare this with a numerically solved solution.

```
>  g1:=plot(y1,x=0..3,style=POINT,symbol=CROSS):
```

To solve numerically we use dsolve with the flag numeric.

```
>  soln1:=dsolve({ode1,y(0)=1,D(y)(0)=0},y(x),numeric);
```

$$soln1 := \mathbf{proc}(\mathit{rkf45\_x})\ldots\mathbf{end\ proc}$$

The numerical solution is plotted and the two compared with display. The Taylor solution is seen to be quite accurate up to $x = 2$ and then the polynomial starts to blow up.

```
>  g2:=odeplot(soln1,[x,y(x)],0..3,numpoints=50):
>  display({g1,g2});
```

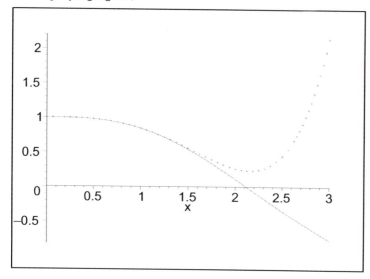

To compute the second solution by the Taylor method we set the coefficient to satisfy the second initial condition and then repeat the previous steps.

Again the Taylor series solution compares well with the numerical solution until $x = 2$ and then it starts to blow up.

```
> p2:=subs({a(0)=0,a(1)=1},p);
```

$$p2 := 2\,a(2) + 6\,a(3)\,x + (12\,a(4) + 1)\,x^2 + (20\,a(5) + a(2))\,x^3 +$$

$$(-\frac{1}{6} + 30\,a(6) + a(3))\,x^4 + (a(4) + 42\,a(7) - \frac{1}{6}\,a(2))\,x^5 +$$

$$(56\,a(8) + \frac{1}{120} + a(5) - \frac{1}{6}\,a(3))\,x^6 +$$

$$(-\frac{1}{6}\,a(4) + a(6) + \frac{1}{120}\,a(2))\,x^7 + O(x^8)$$

```
> for n from 0 to 8 do eq(n):=coeff(p2,x,n) od:
> l:=solve({eq(0),eq(1),eq(2),eq(3),eq(4),eq(5)},
> {a(2),a(3),a(4),a(5),a(6),a(7)});
```

$$l := \left\{ a\,(2) = 0, a\,(3) = 0, a\,(4) = -1/12, a\,(5) = 0, a\,(6) = \frac{1}{180}, a\,(7) = \frac{1}{504} \right\}$$

```
> y2:=subs({a(6) = 1/180, a(2) = 0, a(4) =
> -1/12, a(5) = 0, a(7) =
> 1/504, a(3) = 0},x+sum(a(k)*x^k,k=2..7));
```

$$y2 := x - \frac{1}{12}\,x^4 + \frac{1}{180}\,x^6 + \frac{1}{504}\,x^7$$

```
> g3:=plot(y2,x=0..3,style=POINT,symbol=CIRCLE):
> soln2:=dsolve({ode1,y(0)=0,D(y)(0)=1},y(x),numeric);
```

$$soln2 := \textbf{proc}(rkf45\_x) \ldots \textbf{end proc}$$

```
> g4:=odeplot(soln2,[x,y(x)],0..3,numpoints=50):
> display({g3,g4});
```

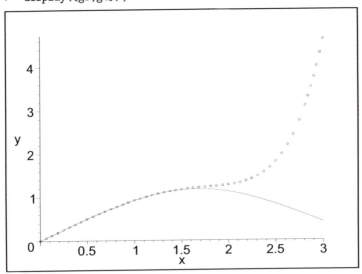

8.1 Nonlinear Differential Equations

The Taylor series method may also be used on non-linear equations. To show how this may be done let us consider the example

$$\frac{d^2y(x)}{dx^2} = x + y^2(x), \text{ with } y(0) = y_0, y'(0) = y_1.$$

We input the right-hand-side as the function $f(x, y)$

```
>   f:= x+y^2;
```

$$f := (x, y) \rightarrow x + y^2$$

```
>   deq1:=diff(y(x),x$2)
>   -f(x,y(x))=0;
```

$$deq1 := (\frac{\partial^2}{\partial x^2} y(x)) - x - y(x)^2 = 0$$

Then as in the linear case we seek a Taylor series solution in the form

$$y(x) = \sum_{k=0}^{\infty} a_k x^k.$$

Notice that the first two coefficients are known from the initial conditions to be $a_0 = y_0$ and $a_1 = y_1$.

```
>   subs(y(x)=
>   sum(a(k)*x^k,k=0..10),deq1);
```

$$(\frac{\partial^2}{\partial x^2} (a(0) + a(1) x + a(2) x^2 + a(3) x^3 + a(4) x^4 + a(5) x^5 + a(6) x^6 + a(7) x^7 + a(8) x^8$$
$$+ a(9) x^9 + a(10) x^{10})) - x - (a(0) + a(1) x + a(2) x^2 + a(3) x^3 + a(4) x^4 + a(5) x^5$$
$$+ a(6) x^6 + a(7) x^7 + a(8) x^8 + a(9) x^9 + a(10) x^{10})^2 = 0$$

In order to take the derivatives we simplify

```
>   P:=expand(%);
```

$$P := -x - 2\,a(1)\,x^3\,a(2) + 2\,a(2) - a(0)^2 + 6\,a(3)\,x - a(1)^2\,x^2 - a(2)^2\,x^4 - a(3)^2\,x^6 - a(4)^2\,x^8$$
$$- a(5)^2\,x^{10} - a(6)^2\,x^{12} - a(8)^2\,x^{16} - a(9)^2\,x^{18} - a(10)^2\,x^{20} - a(7)^2\,x^{14} + 12\,a(4)\,x^2$$
$$+ 20\,a(5)\,x^3 + 30\,a(6)\,x^4 + 42\,a(7)\,x^5 + 56\,a(8)\,x^6 + 72\,a(9)\,x^7 + 90\,a(10)\,x^8$$
$$- 2\,a(1)\,x^4\,a(3) - 2\,a(7)\,x^{15}\,a(8) - 2\,a(7)\,x^{16}\,a(9) - 2\,a(7)\,x^{17}\,a(10) - 2\,a(8)\,x^{17}\,a(9)$$
$$- 2\,a(8)\,x^{18}\,a(10) - 2\,a(1)\,x^5\,a(4) - 2\,a(1)\,x^6\,a(5) - 2\,a(1)\,x^7\,a(6) - 2\,a(1)\,x^8\,a(7)$$
$$- 2\,a(1)\,x^9\,a(8) - 2\,a(1)\,x^{10}\,a(9) - 2\,a(1)\,x^{11}\,a(10) - 2\,a(2)\,x^5\,a(3) - 2\,a(2)\,x^6\,a(4)$$
$$- 2\,a(2)\,x^7\,a(5) - 2\,a(2)\,x^8\,a(6) - 2\,a(2)\,x^9\,a(7) - 2\,a(2)\,x^{10}\,a(8) - 2\,a(2)\,x^{11}\,a(9)$$
$$- 2\,a(2)\,x^{12}\,a(10) - 2\,a(3)\,x^7\,a(4) - 2\,a(3)\,x^8\,a(5) - 2\,a(3)\,x^9\,a(6) - 2\,a(3)\,x^{10}\,a(7)$$
$$- 2\,a(3)\,x^{11}\,a(8) - 2\,a(3)\,x^{12}\,a(9) - 2\,a(3)\,x^{13}\,a(10) - 2\,a(4)\,x^9\,a(5) - 2\,a(4)\,x^{10}\,a(6)$$
$$- 2\,a(4)\,x^{11}\,a(7) - 2\,a(4)\,x^{12}\,a(8) - 2\,a(4)\,x^{13}\,a(9) - 2\,a(4)\,x^{14}\,a(10) - 2\,a(5)\,x^{11}\,a(6)$$
$$- 2\,a(5)\,x^{12}\,a(7) - 2\,a(5)\,x^{13}\,a(8) - 2\,a(5)\,x^{14}\,a(9) - 2\,a(5)\,x^{15}\,a(10) - 2\,a(6)\,x^{13}\,a(7)$$
$$- 2\,a(6)\,x^{14}\,a(8) - 2\,a(6)\,x^{15}\,a(9) - 2\,a(6)\,x^{16}\,a(10) - 2\,a(9)\,x^{19}\,a(10) - 2\,a(0)\,a(1)\,x$$
$$- 2\,a(0)\,a(2)\,x^2 - 2\,a(0)\,a(3)\,x^3 - 2\,a(0)\,a(4)\,x^4 - 2\,a(0)\,a(5)\,x^5 - 2\,a(0)\,a(6)\,x^6$$
$$- 2\,a(0)\,a(7)\,x^7 - 2\,a(0)\,a(8)\,x^8 - 2\,a(0)\,a(9)\,x^9 - 2\,a(0)\,a(10)\,x^{10} = 0$$

We organize our output in terms of powers of x by using `collect`.

```
>  p:= lhs(collect(P,x));
```

$$p := -a(10)^2\,x^{20} - 2\,a(9)\,x^{19}\,a(10) + (-2\,a(8)\,a(10) - a(9)^2)\,x^{18}$$
$$+ (-2\,a(8)\,a(9) - 2\,a(7)\,a(10))\,x^{17} + (-a(8)^2 - 2\,a(7)\,a(9) - 2\,a(6)\,a(10))\,x^{16}$$
$$+ (-2\,a(5)\,a(10) - 2\,a(7)\,a(8) - 2\,a(6)\,a(9))\,x^{15}$$
$$+ (-2\,a(6)\,a(8) - 2\,a(4)\,a(10) - 2\,a(5)\,a(9) - a(7)^2)\,x^{14}$$
$$+ (-2\,a(5)\,a(8) - 2\,a(4)\,a(9) - 2\,a(6)\,a(7) - 2\,a(3)\,a(10))\,x^{13}$$
$$+ (-2\,a(5)\,a(7) - 2\,a(2)\,a(10) - 2\,a(3)\,a(9) - 2\,a(4)\,a(8) - a(6)^2)\,x^{12}$$
$$+ (-2\,a(1)\,a(10) - 2\,a(5)\,a(6) - 2\,a(3)\,a(8) - 2\,a(4)\,a(7) - 2\,a(2)\,a(9))\,x^{11}$$
$$+ (-a(5)^2 - 2\,a(1)\,a(9) - 2\,a(4)\,a(6) - 2\,a(0)\,a(10) - 2\,a(3)\,a(7) - 2\,a(2)\,a(8))\,x^{10}$$
$$+ (-2\,a(3)\,a(6) - 2\,a(0)\,a(9) - 2\,a(2)\,a(7) - 2\,a(4)\,a(5) - 2\,a(1)\,a(8))\,x^9$$
$$+ (-2\,a(0)\,a(8) - 2\,a(1)\,a(7) - 2\,a(3)\,a(5) - a(4)^2 - 2\,a(2)\,a(6) + 90\,a(10))\,x^8$$
$$+ (-2\,a(1)\,a(6) - 2\,a(3)\,a(4) + 72\,a(9) - 2\,a(2)\,a(5) - 2\,a(0)\,a(7))\,x^7$$
$$+ (-2\,a(2)\,a(4) - 2\,a(0)\,a(6) + 56\,a(8) - a(3)^2 - 2\,a(1)\,a(5))\,x^6$$
$$+ (-2\,a(1)\,a(4) + 42\,a(7) - 2\,a(0)\,a(5) - 2\,a(2)\,a(3))\,x^5$$
$$+ (-a(2)^2 + 30\,a(6) - 2\,a(1)\,a(3) - 2\,a(0)\,a(4))\,x^4$$
$$+ (-2\,a(1)\,a(2) - 2\,a(0)\,a(3) + 20\,a(5))\,x^3 + (12\,a(4) - a(1)^2 - 2\,a(0)\,a(2))\,x^2$$
$$+ (6\,a(3) - 2\,a(0)\,a(1) - 1)\,x - a(0)^2 + 2\,a(2)$$

We set the coefficients of the powers of x to zero by collecting these as equations, and then solve for the coefficients in terms of the two **known** coefficients a_0 and a_1.

```
>   for n from
>   0 to 10 do eq(n):=coeff(p,x,n)
>   od;
```

$$eq\,(0) := -\,(a\,(0))^2 + 2\,a\,(2)$$

$$eq\,(1) := -2\,a\,(0)\,a\,(1) + 6\,a\,(3) - 1$$

$$eq\,(2) := -2\,a\,(0)\,a\,(2) - (a\,(1))^2 + 12\,a\,(4)$$

$$eq\,(3) := -2\,a\,(0)\,a\,(3) - 2\,a\,(1)\,a\,(2) + 20\,a\,(5)$$

$$eq\,(4) := -2\,a\,(0)\,a\,(4) - 2\,a\,(1)\,a\,(3) - (a\,(2))^2 + 30\,a\,(6)$$

$$eq\,(5) := -2\,a\,(0)\,a\,(5) - 2\,a\,(1)\,a\,(4) - 2\,a\,(2)\,a\,(3) + 42\,a\,(7)$$

$$eq\,(6) := -2\,a\,(0)\,a\,(6) - 2\,a\,(1)\,a\,(5) - 2\,a\,(2)\,a\,(4) - (a\,(3))^2 + 56\,a\,(8)$$

$$eq\,(7) := -2\,a\,(0)\,a\,(7) - 2\,a\,(1)\,a\,(6) - 2\,a\,(2)\,a\,(5) - 2\,a\,(3)\,a\,(4) + 72\,a\,(9)$$

$$eq\,(8) := -2\,a\,(0)\,a\,(8) - 2\,a\,(1)\,a\,(7) - 2\,a\,(2)\,a\,(6) - 2\,a\,(3)\,a\,(5) - (a\,(4))^2 + 90\,a\,(10)$$

$$eq\,(9) := -2\,a\,(0)\,a\,(9) - 2\,a\,(1)\,a\,(8) - 2\,a\,(2)\,a\,(7) - 2\,a\,(3)\,a\,(6) - 2\,a\,(4)\,a\,(5)$$

$$eq\,(10) := -2\,a\,(1)\,a\,(9) - 2\,a\,(0)\,a\,(10) - 2\,a\,(2)\,a\,(8) - 2\,a\,(3)\,a\,(7) - 2\,a\,(4)\,a\,(6) - (a\,(5))^2$$

```
>   solve({eq(0),eq(1),
>   eq(2),eq(3),eq(4),eq(5)},
>   {a(2),a(3),a(4),a(5),a(6),a(7)});
```

$$\{a\,(2) = 1/2\,(a\,(0))^2\,,a\,(3) = 1/3\,a\,(0)\,a\,(1) + 1/6, a\,(4) = 1/12\,(a\,(0))^3 + 1/12\,(a\,(1))^2$$

$$a\,(5) = 1/12\,a\,(1)\,(a\,(0))^2 + \frac{a\,(0)}{60}, a\,(6) = \frac{(a\,(0))^4}{72} + 1/36\,a\,(0)\,(a\,(1))^2 + \frac{a\,(1)}{90},$$

$$a\,(7) = \frac{a\,(1)\,(a\,(0))^3}{63} + \frac{(a\,(1))^3}{252} + \frac{(a\,(0))^2}{210}\}$$

The truncated series solution to the differential equation is thereby given as

$$a_0 + a_1 x + \frac{x^2}{2}a_0^2 + (\frac{1}{3}a_0 a_1 + \frac{1}{6})x^3$$

$$\frac{1}{12}(a_1^2 + a_0^3)x^4 + (\frac{1}{12}a_0^2 a_1 + \frac{1}{60}a_0)x^5$$

$$(\frac{1}{72}a_0^4 + \frac{1}{36}a_0 a_1^2 + \frac{1}{90}rm_1)x^6 +$$

$$(\frac{1}{252}a_1^3 + \frac{1}{63}a_0^3 a_1 + \frac{1}{210}rm_0^2)x^7 + \cdots$$

8.2   Regular-Singular Points

Our next example of how to solve an equation with analytical coefficients involves finding a solution about a **regular-singular** point [12], Chapter 11. We notice if we divide the next differential equation by x^2, the coefficient of the leading derivative is 1, and the coefficients of the other terms are singular at $x = 0$. Moreover, the orders are such that the coefficients of the first-order derivative and also the unknown have first-order singularities.

```
>   ode2:= x^2*diff(y(x),
>   x$2)+x*diff(y(x),x)+
>   2*x*y(x);
```

$$ode2 := x^2 \left(\frac{\partial^2}{\partial x^2} y(x)\right) + x \left(\frac{\partial}{\partial x} y(x)\right) + 2\,x\,y(x)$$

For expansions about regular singular points we expect power series of the form

$$y(x) = \sum_{k=0}^{\infty} a_k (x - x_0)^{k+m};$$

about a regular singular point st $x = x_0$. Hence, for our case we substitute in a truncated series

$$y(x) = \sum_{k=0}^{N} a_k x^{k+m};$$

$p := 2\,x^{(m+1)}\,a(0) + x^{(m+3)}\,m^2\,a(3) + x^m\,m^2\,a(0) + 16\,x^{(m+4)}\,a(4) + 2\,x^{(m+1)}\,m\,a(1)$

$\quad + x^{(m+6)}\,m^2\,a(6) + x^{(m+5)}\,m^2\,a(5) + 10\,x^{(m+5)}\,m\,a(5) + 4\,x^{(m+2)}\,m\,a(2)$

$\quad + 16\,x^{(m+8)}\,m\,a(8) + 8\,x^{(m+4)}\,m\,a(4) + 14\,x^{(m+7)}\,m\,a(7) + x^{(m+2)}\,m^2\,a(2)$

$\quad + x^{(m+4)}\,m^2\,a(4) + x^{(m+1)}\,m^2\,a(1) + x^{(m+7)}\,m^2\,a(7) + 6\,x^{(m+3)}\,m\,a(3) + 9\,x^{(m+3)}\,a(3)$

$\quad + 25\,x^{(m+5)}\,a(5) + 36\,x^{(m+6)}\,a(6) + 49\,x^{(m+7)}\,a(7) + 64\,x^{(m+8)}\,a(8) + 2\,x^{(m+4)}\,a(3)$

$\quad + 2\,x^{(m+5)}\,a(4) + 2\,x^{(m+6)}\,a(5) + 2\,x^{(m+7)}\,a(6) + 4\,x^{(m+2)}\,a(2) + 2\,x^{(m+8)}\,a(7)$

$\quad + 2\,x^{(m+9)}\,a(8) + x^{(m+1)}\,a(1) + x^{(m+8)}\,m^2\,a(8) + 2\,x^{(m+2)}\,a(1) + 12\,x^{(m+6)}\,m\,a(6)$

$\quad + 2\,x^{(m+3)}\,a(2)$

It is convenient to first factor out the term x^m before collecting coefficients.

```
>   q(x):=simplify(p/(x^m));
```

$q(x) := x^6\, m^2\, a(6) + 10\, x^5\, m\, a(5) + 4\, x^2\, m\, a(2) + 16\, x^8\, m\, a(8) + 2\, x^3\, a(2) + 25\, a(5)\, x^5$
$\quad + x^7\, m^2\, a(7) + x^5\, m^2\, a(5) + x^8\, m^2\, a(8) + 12\, x^6\, m\, a(6) + 14\, x^7\, m\, a(7) + 2\, x^5\, a(4)$
$\quad + x^3\, m^2\, a(3) + 2\, x\, m\, a(1) + 8\, x^4\, m\, a(4) + 2\, x^8\, a(7) + x^2\, m^2\, a(2) + 6\, x^3\, m\, a(3)$
$\quad + x\, m^2\, a(1) + 2\, x^7\, a(6) + x^4\, m^2\, a(4) + 16\, x^4\, a(4) + x\, a(1) + 2\, x\, a(0) + 2\, x^6\, a(5)$
$\quad + 4\, x^2\, a(2) + 2\, x^9\, a(8) + 9\, x^3\, a(3) + 36\, x^6\, a(6) + 49\, x^7\, a(7) + 64\, x^8\, a(8) + 2\, x^4\, a(3)$
$\quad + m^2\, a(0) + 2\, x^2\, a(1)$

Now we collect our equations as before using `coeff`.

```
>   for n from 0 to 8 do
>   equ(n):=coeff(q(x),x,n);
>   od;
```

$$\mathrm{equ}(0) := m^2\, \mathrm{a}(0)$$

$$equ\,(1) \;:=\; (1+m)^2\, a\,(1) + 2\, a\,(0)$$

$$equ\,(2) \;:=\; 2\, a\,(1) + (2+m)^2\, a\,(2)$$

$$equ\,(3) \;:=\; 2\, a\,(2) + (3+m)^2\, a\,(3)$$

$$equ\,(4) \;:=\; 2\, a\,(3) + (4+m)^2\, a\,(4)$$

$$equ\,(5) \;:=\; 2\, a\,(4) + (5+m)^2\, a\,(5)$$

$$equ\,(6) \;:=\; 2\, a\,(5) + (6+m)^2\, a\,(6)$$

$$equ\,(7) \;:=\; 2\, a\,(6) + (7+m)^2\, a\,(7)$$

$$equ\,(8) \;:=\; 2\, a\,(7) + (8+m)^2\, a\,(8)$$

From equation (1), namely $m^2\, a_0 = 0$, we get the **indicial equation** $m^2 = 0$; hence, $m = 0$. We substitute this value for the index m into each of the equations and solve for the coefficients of the power series.

```
>   for n from 0 to 8 do
>   equ(n):=subs(m=0,equ(n)); od;
```

$$\mathrm{equ}(0) := 0$$

$$\mathrm{equ}(1) := \mathrm{a}(1) + 2\,\mathrm{a}(0)$$

$$\mathrm{equ}(2) := 2\,\mathrm{a}(1) + 4\,\mathrm{a}(2)$$

$$\mathrm{equ}(3) := 2\,\mathrm{a}(2) + 9\,\mathrm{a}(3)$$

$$\mathrm{equ}(4) := 2\,\mathrm{a}(3) + 16\,\mathrm{a}(4)$$

$$\mathrm{equ}(5) := 2\,\mathrm{a}(4) + 25\,\mathrm{a}(5)$$

$$\mathrm{equ}(6) := 2\,\mathrm{a}(5) + 36\,\mathrm{a}(6)$$

$$\text{equ}(7) := 2\,a(6) + 49\,a(7)$$

$$\text{equ}(8) := 2\,a(7) + 64\,a(8)$$

```
>   solve({equ(1),equ(2)
>   ,equ(3),equ(4),equ(5),
>   equ(6),equ(7),equ(8)},
>   {a(1), a(2), a(3), a(4), a(5), a(6), a(7), a(8)})
```

$$\{a(2) = 0,\ a(3) = -\frac{1}{6}\,a(0),\ a(4) = \frac{9}{392}\,a(0),\ a(6) = \frac{71}{17640}\,a(0),\ a(7) = -\frac{23}{30870}\,a(0),$$

$$a(8) = -\frac{1193}{1975680}\,a(0),\ a(1) = -\frac{27}{98}\,a(0),\ a(5) = \frac{1}{120}\,a(0)\}$$

8.3   Programs for Finding Solutions

We next consider writing programs to automatically compute the series solutions of arbitrary equations of the form

$$x^2 \frac{d^2\,y(x)}{dx^2} + x^2 p(x) \frac{d\,y(x)}{dx} + x^2\,q(x)y(x) = 0,$$

where

$$x\,p(x) = \sum_{n=0}^{\infty} p_n\,x^n,\quad x^2\,q(x) = \sum_{n=0}^{\infty} q_n\,x^n,$$

both converge in an interval about the origin. The Frobenius method is to seek a solution in the form

$$y(x) = \sum_{n=0}^{\infty} a_n\,x^{n+m}$$

where the index m is to be determined and the coefficient $a_0 = 1$. The index m satisfies the indicial equation

$$m(m-1) + p_0 m + q_0 = 0.$$

We start with the differential equation

$$\frac{d^2 y(x)}{dx^2} - \frac{d\,y(x)}{dx} + \cos(x)\,y(x)$$

. then the coefficients are $p(x) := -1$, and $q(x) := \cos(x)$. First expand the coefficients using `taylor` and then `convert` the Taylor series to polynomials using

```
>   tp:=convert(taylor(-1,x=0,10),polynom);
>   tq:=convert(taylor(cos(x),x=0,10),polynom);
```

$$tp := -1$$

$$tq := 1 - \frac{1}{2} x^2 + \frac{1}{24} x^4 - \frac{1}{720} x^6 + \frac{1}{40320} x^8$$

Next we want to extract the coefficients of the Taylor polynomials tp and tq. This is done using coeff for all terms x^j, $j = 1, \ldots$. However, the constant term has to be extracted differently. We use subs for that. We use seq to form a sequence of the Taylor coefficients and these are then printed.

```
>   p[0]:=subs(x=0,tp); q[0]:=subs(x=0,tq);
>   for j from 1 to 10 do p[j] := coeff(tp,x^(j)) od:
>   Lp:=seq(p||l=p[l],l=0..10); for k from 1 to 10 do q[k] :=
>   coeff(tq,x^(k)) od: Lq:=seq(p||m=q[m],m=0..10);
```

$$p_0 := -1$$

$$q_0 := 1$$

$$Lp := p0 = -1,\ p1 = 0,\ p2 = 0,\ p3 = 0,\ p4 = 0,\ p5 = 0,\ p6 = 0,\ p7 = 0,$$
$$p8 = 0,\ p9 = 0,\ p10 = 0$$

$$Lq := p0 = 1,\ p1 = 0,\ p2 = \frac{-1}{2},\ p3 = 0,\ p4 = \frac{1}{24},\ p5 = 0,$$
$$p6 = \frac{-1}{720},\ p7 = 0,\ p8 = \frac{1}{40320},\ p9 = 0,\ p10 = 0$$

Next we want to extract the indicial equation. This is done using the well known formula for $f(x)$

$$f(x) := x^2 + (p_0 - 1) + q_0.$$

```
>   f(x):=x^2+(p[0]-1)*x+q[0];
```

$$f(x) := x^2 - 2x + 1$$

We now obtain the roots of the indicial equation, which we denote as the pair $r := [r_1, r_2]$

```
>   r:=[solve(f(x),x)];
```

$$r := [1, 1]$$

a repeated root in this case.

```
>   m:=r[1];
```

$$m := 1$$

Let us check whether we can substitute values into $f(x)$. We could use evalf here, but we used subs.

```
>   subs(x=5+m,f(x));
```

25

Now let us compute the first 10 coefficients in the series expansion

$$y_1(x) := x^m \sum_{n=0}^{\infty} a_n x^n$$

This is done using the recursion formula

$$a_n := -\frac{1}{f(n+m)} \sum_{k=0}^{n-1} ((k+m)p_{n-k} + q_{n-k} a_k)$$

For the example we started with we obtain

```
> a[0]:=1; for r from 1 to 10 do a[r] :=-
> sum(((s+m)*p[r-s]+ q[r-s])*a[s],s=0..r-1)/subs(x=r+m,f(x)) od:
> L:=seq(a||l=a[l],l=0..10);
```

$$a_0 := 1$$

$$L := a0 = 1,\ a1 = 0,\ a2 = \frac{1}{8},\ a3 = 0,\ a4 = \frac{1}{768},\ a5 = 0,\ a6 = \frac{-73}{829440},\ a7 = 0,$$

$$a8 = \frac{587}{743178240},\ a9 = 0,\ a10 = \frac{12353}{445906944000}$$

We may then write down a truncated solution to the differential equation as

```
> solution :=sum(a[l]*x^(m+l),l=1..10);
```

$$solution := \frac{1}{8} x^3 + \frac{1}{768} x^5 - \frac{73}{829440} x^7 + \frac{587}{743178240} x^9 + \frac{12353}{445906944000} x^{11}$$

Having seen how to construct a solution line by line let us attempt to write a program which does this. As this will not be so easy, we begin by partitioning the problem into sections which will each perform part of the task. The first part is a mini-program to construct the Taylor coefficients of the coefficients $p(x)$ and $q(x)$.

```
> coeffs_p_q := proc(p,q,n) description
> "This program expands the coefficients of the differential
> equation as a Taylor series"; local tp, tq, j, k,m, t,P,Q,Lp,Lq;
> tp:=convert(taylor(p,x=0,n),polynom);
> tq:=convert(taylor(q,x=0,n),polynom); P[0]:=subs(x=0,tp);
> Q[0]:=subs(x=0,tq); for j from 1 to n do P[j] := coeff(tp,x^(j))
> od:for k from 1 to n do Q[k] := coeff(tq,x^(k)) od:
> Lp:=seq(P||m=P[m],m=0..n);Lq:=seq(Q||l=Q[l],l=0..n);
> print([Lp,Lq]);end;
```

$coeffs\_p\_q := \mathbf{proc}(p, q, n)$
$\mathbf{local}\ tp,\ tq,\ j,\ k,\ m,\ t,\ P,\ Q,\ Lp,\ Lq;$
$\mathbf{description}$ "This program expands the coefficients of the differential equation as a Taylor series";
$\quad tp := \mathrm{convert}(\mathrm{taylor}(p,\ x = 0,\ n),\ polynom);$
$\quad tq := \mathrm{convert}(\mathrm{taylor}(q,\ x = 0,\ n),\ polynom);$
$\quad P_0 := \mathrm{subs}(x = 0,\ tp);$
$\quad Q_0 := \mathrm{subs}(x = 0,\ tq);$
$\quad \mathbf{for}\ j\ \mathbf{to}\ n\ \mathbf{do}\ P_j := \mathrm{coeff}(tp,\ x^j)\ \mathbf{end\ do};$
$\quad \mathbf{for}\ k\ \mathbf{to}\ n\ \mathbf{do}\ Q_k := \mathrm{coeff}(tq,\ x^k)\ \mathbf{end\ do};$
$\quad Lp := \mathrm{seq}(P\|m = P_m,\ m = 0..n);$
$\quad Lq := \mathrm{seq}(Q\|l = Q_l,\ l = 0..n);$
$\quad \mathrm{print}([Lp,\ Lq])$
$\mathbf{end\ proc}$

```
>  coeffs_p_q(-1,cos(x),5);
```

$[P0 = -1,\ P1 = 0,\ P2 = 0,\ P3 = 0,\ P4 = 0,\ P5 = 0,\ Q0 = 1,\ Q1 = 0,$
$Q2 = \dfrac{-1}{2},\ Q3 = 0,\ Q4 = \dfrac{1}{24},\ Q5 = 0]$

```
>  frobenius_1 :=proc(p,q,n,x) local k, j,
>  tp, tq, r,P,Q,t,m,a,b,l,L; description "This program computes the
>  indicial equation, finds the indices, chooses an index, and
>  computes the first n coefficients of the series expansion";
>  tp:=convert(taylor(p,x=0,n),polynom);
>  tq:=convert(taylor(q,x=0,n),polynom); P[0]:=subs(x=0,tp);
>  Q[0]:=subs(x=0,tq); for j from 1 to n do P[j] := coeff(tp,x^(j))
>  od;for k from 1 to n do Q[k] := coeff(tq,x^(k)) od;
>  f(x):=x^2+(P[0]-1)*x+Q[0]; print("indicial equation"
>  ,f(x));r:=[solve(f(x)=0,x)]; print("indices" ,r);m:=r[1]; a[0]:=1;
>  for b from 1 to n do a[b] :=- sum(((c+m)*P[b-c]+
>  Q[b-c])*a[c],c=0..b-1)/subs(x=b+m,f(x)) od; L:=seq(a||l=a[l],l= 1
>  .. n );
>  print("coefficients",L); end proc;
```

frobenius__1 := **proc**(p, q, n, x)
local $k, j, tp, tq, r, P, Q, t, m, a, b, l, L;$
description "This program computes the indicial equation,
finds the indices, chooses an index, and computes the first n
coefficients of the series expansion";

$tp := $ convert$($taylor$(p, x = 0, n),$ *polynom*$)\,;$
$tq := $ convert$($taylor$(q, x = 0, n),$ *polynom*$)\,;$
$P_0 := $ subs$(x = 0, tp)\,;$
$Q_0 := $ subs$(x = 0, tq)\,;$
for j **to** n **do** $P_j := $ coeff(tp, x^j) **end do**$;$
for k **to** n **do** $Q_k := $ coeff(tq, x^k) **end do**$;$
f$(x) := x^2 + (P_0 - 1) * x + Q_0\,;$
print$($"indicial equation", f$(x))\,;$
$r := [$solve$($f$(x) = 0, x)]\,;$
print$($"indices", $r)\,;$
$m := r_1\,;$
$a_0 := 1\,;$
for b **to** n **do** $a_b :=$
$- $ sum$(((c + m) * P_{b-c} + Q_{b-c}) * a_c, c = 0..b - 1)/$
subs$(x = b + m,$ f$(x))$
end do$;$
$L := $ seq$(a||l = a_l, l = 1..n)\,;$
print$($"coefficients", $L)$
end proc

> `frobenius_1(-1,cos(x),5,x);`

"indicial equation", $x^2 - 2x + 1$

"indices", $[1, 1]$

"coefficients", $a1 = 0, a2 = \dfrac{1}{8}, a3 = 0, a4 = \dfrac{1}{768}, a5 = 0$

If the two roots of the indicial equation do not differ by an integer then
we may find solutions in the form

$$y_1(x) = x^{m_1}\left(1 + \sum_{n=1}^{\infty} a_n x^n\right)$$

$$y_2(x) = x^{m_2}\left(1 + \sum_{n=1}^{\infty} b_n x^n\right)$$

However, if the two indices differ by an integer we may not get a second
solution in this way. If the two indices are the same, i.e. a double root, then

we do **NOT** get a second solution in this form! Hence, the choice of the index we use to construct the first solution must carefully chosen. In the case where roots differ by an integer, the root with the larger real part always gives rise to a solution. Hence, we wish to first construct this solution. We make use of the command `type`. The command `type` allows us to check whether an expression is of a certain "type". One such "type" is `integer`. A particular cases which must be distinguished is when the difference between the two roots of the indicial equation differ by an integer. A subcategory which is of further interest occurs when the roots are identical, i.e. they differ by zero. `type` returns either `true` or `false`. Let us experiment a bit with `type`

```
>   type(5,integer);
```

$$true$$

```
>   type(Pi,integer);
```

$$false$$

We recall our example from the previous session

```
>   tp:=convert(taylor(-1,x=0,10),polynom):tq:=co
>   nvert(taylor(cos(x),x=0,10),polynom):
```

```
>   f(x):=x^2+(p[0]-1)*x+q[0];
```

$$f(x) := x^2 - 2x + 1$$

```
>   r:=[solve(f(x),x)];
```

$$r := [1, 1]$$

```
>   m.1:=r[1]; m.2:=r[2];
```

$$m1 := 1$$

$$m2 := 1$$

Let us write a mini-program to choose one of the roots. For the moment we consider only the possibility of *real* roots. `pos` is designed to pick the larger of two real roots. Once we understand how to incorporate the choosing mechanism into `frobenius` we will consider the case of complex roots.

```
>   type(m.1-m.2,integer);
```

$$true$$

```
>   pos := proc(m) if type(
>   m[1]-m[2],integer) then if m[1] >m[2] then  m[1] else m[2]
>   fi;else m[1] fi: end;
```

$pos := \mathbf{proc}(m)$

    if $\text{type}(m_1 - m_2,\ integer)$ **then if** $m_2 < m_1$ **then** m_1 **else** m_2 **fi else** m_1 **fi**

  end

```
>  pos([2,3]);
```

$$3$$

Notice the pos actually can make a choice between two complex roots; however, it may make the wrong choice for our purposes. More about this shortly!

```
>  pos([2+I*4,Pi+I*4]);
```

$$2 + 4\,I$$

Now we write a new frobenius program incorporating the "choice"

```
>  frobenius_2 :=proc(p,q,n,x) description
>  "This program computes the power series solution to an equation
>  with a regular singular point at the origin. It first solves the
>  indicial equation and then computes the coefficients of the power
>  series respective to a particular index" ;local k, j, tp, tq,
>  r,P,Q,a,b,m,l,L; tp:=convert(taylor(p,x=0,n),polynom);
>  tq:=convert(taylor(q,x=0,n),polynom); P[0]:=subs(x=0,tp);
>  Q[0]:=subs(x=0,tq); for j from 1 to n do P[j] := coeff(tp,x^(j))
>  od;for k from 1 to n do Q[k] := coeff(tq,x^(k)) od;
>  f(x):=x^2+(P[0]-1)*x+Q[0]; print(f(x)); r:=[solve(f(x)=0,x)]:
>  print("indices are ",[r[1],r[2]]); if type( r[1]-r[2],integer)
>  then if r[1] >r[2] then  m:=r[1] else m:=r[2] fi;else m:=r[1] fi;
>  print("index chosen  ",m);
>  a[0]:=1;
>  for b from 1 to n do a[b]  :=- sum(((c+m)*P[b-c]+
>  Q[b-c])*a[c],c=0..b-1)/subs(x=b+m,f(x)) od: for l from 0 to n do
>  a[l] od: L:=seq(a||h=a[h],h=0..n);print("expansion coefficints
>  ",L);end;
```

```
frobenius_2 := proc(p, q, n, x)
local k, j, tp, tq, r, P, Q, a, b, m, l, L;
description "This program computes the power series solution to
an equation with a regular singular point at the origin. It first
solves the indicial equation and then computes the coefficients
of the power series respective to a particular index";
    tp := convert(taylor(p, x = 0, n), polynom);
    tq := convert(taylor(q, x = 0, n), polynom);
    P₀ := subs(x = 0, tp);
    Q₀ := subs(x = 0, tq);
    for j to n do Pⱼ := coeff(tp, xʲ) end do;
    for k to n do Qₖ := coeff(tq, xᵏ) end do;
    f(x) := x² + (P₀ − 1) * x + Q₀;
    print(f(x));
    r := [solve(f(x) = 0, x)];
    print("indices are ", [r₁, r₂]);
    if type(r₁ − r₂, integer) then if r₂ < r₁ then m := r₁ else m := r₂
    end if
    else m := r₁
    end if;
    print("index chosen ", m);
    a₀ := 1;
    for b to n do aᵦ :=
        −sum(((c + m) * P_{b−c} + Q_{b−c}) * a_c, c = 0..b − 1)/
    subs(x = b + m, f(x))
    end do;
    for l from 0 to n do aₗ end do;
    L := seq(a||h = aₕ, h = 0..n);
    print("expansion coefficints ", L)
end proc
```

Again we check this program on our example.

```
>   frobenius_2(-1,cos(x),5,x);
```

$$x^2 - 2x + 1$$

"indices are ", $[1, 1]$

"index chosen ", 1

"expansion coefficints ", $a0 = 1$, $a1 = 0$, $a2 = \dfrac{1}{8}$, $a3 = 0$, $a4 = \dfrac{1}{768}$, $a5 = 0$

Let us now choose coefficients which lead to complex roots and see what frobenius_2 does.

```
>   frobenius_2(1,cos(x),5,x);
```

$$x^2 + 1$$

"indices are ", $[I, -I]$

"index chosen ", I

"expansion coefficints ", $a0 = 1$, $a1 = 0$, $a2 = \dfrac{1}{16} - \dfrac{1}{16} I$, $a3 = 0$,

$$a4 = \frac{-1}{768} - \frac{1}{768} I, \; a5 = 0$$

frobenius_2 appears to work fine. Check it against some cases where you know the solution and *benchmark* it!

```
>   frobenius_3 :=proc(p,q,n,x) description "
>   This program uses the Frobenious method. To decide which index to
>   use it takes the one with larger real part. The indicial equation
>   is printed, its solution, the index chosen, and the first n
>   coefficients in decimal form are printed." ;local k, j, tp, tq,
>   r,P,Q,a,b,m,l,L; tp:=convert(taylor(p,x=0,n),polynom);
>   tq:=convert(taylor(q,x=0,n),polynom); P[0]:=subs(x=0,tp);
>   Q[0]:=subs(x=0,tq); for j from 1 to n do P[j] := coeff(tp,x^(j))
>   od;for k from 1 to n do Q[k] := coeff(tq,x^(k)) od;
>   f(x):=x^2+(P[0]-1)*x+Q[0]; print(f(x)); r:=[solve(f(x)=0,x)]:
>   print("idices are ",[r[1],r[2]]); if type( r[1]-r[2],integer) then
>   if r[1] >r[2] then  m:=r[2] else m:=r[1] fi;elif
>   type( r[1]-r[2],positive) then m:=r[1]
>   else m:=r[2] fi: print("index chosen ",m);
>   a[0]:=1;
>   for b from 1 to n do a[b] :=- sum(((c+m)*P[b-c]+
>   Q[b-c])*a[c],c=0..b-1)/subs(x=b+m,f(x)) od:
>   L:=seq(a||l=a[l],l=0..n); print("expansion coefficients are " ,L);
>   end;
```

$frobenius\_3 := \mathbf{proc}(p, q, n, x)$

$\mathbf{local}\, k,\, j,\, tp,\, tq,\, r,\, P,\, Q,\, a,\, b,\, m,\, l,\, L;$

$\mathbf{description}$ "This program uses the Frobenious method.
To decide which index to use it takes the one with larger real part.
The indicial equation is printed, its solution, the index chosen, and
the first n coefficients in decimal form are printed.";

> $tp := \text{convert}(\text{taylor}(p,\, x = 0,\, n),\, polynom)\,;$

> $tq := \text{convert}(\text{taylor}(q,\, x = 0,\, n),\, polynom)\,;$

> $P_0 := \text{subs}(x = 0,\, tp)\,;$

> $Q_0 := \text{subs}(x = 0,\, tq)\,;$

> $\mathbf{for}\, j\, \mathbf{to}\, n\, \mathbf{do}\, P_j := \text{coeff}(tp,\, x^j)\, \mathbf{end\ do}\,;$

> $\mathbf{for}\, k\, \mathbf{to}\, n\, \mathbf{do}\, Q_k := \text{coeff}(tq,\, x^k)\, \mathbf{end\ do}\,;$

> $\text{f}(x) := x^2 + (P_0 - 1) * x + Q_0\,;$

> $\text{print}(\text{f}(x))\,;$

> $r := [\text{solve}(\text{f}(x) = 0,\, x)]\,;$

> $\text{print}(\text{"idices are "},\, [r_1, r_2])\,;$

> $\mathbf{if}\,\text{type}(r_1 - r_2,\, integer)\,\mathbf{then\ if}\,r_2 < r_1\,\mathbf{then}\, m := r_2\,\mathbf{else}\, m := r_1$
$\mathbf{end\ if}$

> $\mathbf{elif}\,\text{type}(r_1 - r_2,\, positive)\,\mathbf{then}\, m := r_1$

> $\mathbf{else}\, m := r_2$

> $\mathbf{end\ if};$

> $\text{print}(\text{"index chosen "},\, m)\,;$

> $a_0 := 1\,;$

> $\mathbf{for}\, b\, \mathbf{to}\, n\, \mathbf{do}\, a_b :=$
$\quad - \text{sum}(((c + m) * P_{b-c} + Q_{b-c}) * a_c,\, c = 0..b - 1)/$
$\quad \text{subs}(x = b + m,\, \text{f}(x))$

> $\mathbf{end\ do};$

> $L := \text{seq}(a||l = a_l,\, l = 0..n)\,;$

> $\text{print}(\text{"expansion coefficients are "},\, L)$

> $\mathbf{end\ proc}$

>   `frobenius_3(1,cos(x),5,x);`

$$x^2 + 1$$

$$\text{"idices are "},\, [I, -I]$$

$$\text{"index chosen "},\, -I$$

$$\text{"expansion coefficients are "},\, a0 = 1,\, a1 = 0,\, a2 = \frac{1}{16} + \frac{1}{16}I,\, a3 = 0,$$

$$a4 = \frac{-1}{768} + \frac{1}{768}I,\, a5 = 0$$

If the root is a double root, i.e. $m_1 = m_2$ then the second solution may be found in the form

$$y_2(x) = y_1(x) \ln(|x|) + |x|^{m_1} \sum_{n=1}^{\infty};$$

whereas, if $m_1 - m_2$ is a positive integer then the second solution is found in the form

$$y_2(x) = a\, y_1(x) \ln(|x|) + |x|^{m_2} \sum_{n=1}^{\infty}.$$

We now consider the case of complex roots and how to choose the root with the larger real part to start with. If the two roots differ by an integer, we need to find our first solution using that root with largest real-part. MAPLE has a command Re which chooses the real part of an expression. The following mini-program chooses the root with the larger real part when the roots differ by an integer. It does not do this if they do not differ by an integer; however, in that instance it does not matter which root it chooses. It is not difficult to change this to always choose the root with the largest real part. Do this!

```
> condition := proc(m) if type(
> m[1]-m[2],integer) then if Re( m[1]) >Re(m[2]) then   m[1]
> else m[2] fi;else   m[2] fi; end;
```

$condition := \mathbf{proc}(m)$

   $\mathbf{if}\ \mathrm{type}(m_1 - m_2,\ integer)\ \mathbf{then\ if}\ \mathrm{Re}\,(m_2) < \mathrm{Re}\,(m_1)\ \mathbf{then}\ m_1\ \mathbf{else}\ m_2\ \mathbf{fi}$

   $\mathbf{else}\ m_2$

   $\mathbf{fi}$

$\mathbf{end}$

We incorporate this selection mechanism into frobenius as follows

```
> frobenius_4 :=proc(p,q,n,x)
> description " This program uses the Frobenious method. To decide
> which index to use it takes the one with larger real part. The
> indicial equation is printed, its solution, the index chosen, and
> the first n coefficients in decimal form are printed." ;local k,
> j, tp, tq, r,P,Q,a,b,m,l,L; tp:=convert(taylor(p,x=0,n),polynom):
> tq:=convert(taylor(q,x=0,n),polynom):P[0]:=subs(x=0,tp);
> Q[0]:=subs(x=0,tq); for j from 1 to n do P[j] := coeff(tp,x^(j))
> od;for k from 1 to n do Q[k] := coeff(tq,x^(k)) od;
> f(x):=x^2+(P[0]-1)*x+Q[0]; print(f(x)); r:=[solve(f(x)=0,x)];
> print("indices are",[r[1],r[2]]); if type( r[1]-r[2],integer) then
> if Re( r[1]) >Re(r[2]) then m:= r[1] else m:=r[2] fi;else   m:=r[2]
> fi; print("index chosen is", m);
> a[0]:=1;
> for b from 1 to n do a[b] :=- sum(((c+m)*P[b-c]+
> Q[b-c])*a[c],c=0..b-1)/subs(x=b+m,f(x)) od:
> L:=seq( a||l=evalf(a[l]),l=0..n); print("coefficients are ",L);end;
```

frobenius_4 := **proc**(*p, q, n, x*)

local *k, j, tp, tq, r, P, Q, a, b, m, l, L*;

description "This program uses the Frobenious method.
To decide which index to use it takes the one with larger real part.
The indicial equation is printed, its solution, the index
chosen, and the first n coefficients in decimal form are printed.";

tp := convert(taylor(p, $x = 0$, n), *polynom*);

tq := convert(taylor(q, $x = 0$, n), *polynom*);

P_0 := subs($x = 0$, tp);

Q_0 := subs($x = 0$, tq);

for j **to** n **do** P_j := coeff(tp, x^j) **end do**;

for k **to** n **do** Q_k := coeff(tq, x^k) **end do**;

f(x) := $x^2 + (P_0 - 1) * x + Q_0$;

print(f(x));

r := [solve(f(x) = 0, x)];

print("indices are", [r_1, r_2]);

if type($r_1 - r_2$, *integer*) **then**

  if Re(r_2) < Re(r_1) **then** $m := r_1$ **else** $m := r_2$ **end if**

else $m := r_2$

end if;

print("index chosen is", m);

$a_0 := 1$;

for b **to** n **do** a_b :=

  $- $sum((($c + m$) * $P_{b-c} + Q_{b-c}$) * a_c, $c = 0..b - 1$)/

subs($x = b + m$, f(x))

end do;

L := seq($a_l || l$ = evalf(a_l), $l = 0..n$);

print("coefficients are ", L)

end proc

We test this program on our old examples

> `frobenius_4(-1,cos(x),5,x);`

$$x^2 - 2x + 1$$

$$\text{"indices are", } [1, 1]$$

$$\text{"index chosen is", } 1$$

$$\text{"coefficients are ", } a0 = 1., a1 = 0., a2 = .1250000000,$$
$$a3 = 0., a4 = .001302083333, a5 = 0.$$

> `frobenius_4(-3,cos(x),5,x);`

$$x^2 - 4x + 1$$

$$\text{"indices are", } [2 + \sqrt{3},\, 2 - \sqrt{3}]$$
$$\text{"index chosen is", } 2 - \sqrt{3}$$

"coefficients are ", $a0 = 1.$, $a1 = 0.$, $a2 = -0.1707531754$, $a3 = 0.$, $a4 = -0.05926648495$, $a5 = 0.$

8.4 Projects

Write a program to choose the second solution when the roots of the indicial equation differ by an integer.

Chapter 9

Nonlinear Autonomous Systems

We revisit the autonomous system in this chapter, in particular, nonlinear, autonomous systems. An example of such a system is provided by the motion of a simple pendulum. This motion can be modelled by the initial-value problem:

$$\theta'' + \frac{g}{\ell} \sin \theta = 0, \quad \theta(0) = \theta_0, \ \theta'(0) = \theta_0'.$$

Here θ is the angular displacement from the vertical, ℓ is the length of the pendulum, and g is the gravitational acceleration. To simplify the computation, without loss of generality, we may assume that the time scale has been normalized so that the governing equation becomes:

(E)
$$\frac{d^2\theta}{dt^2} + \sin \theta = 0, \quad t > 0$$

together with the initial conditions

(I)
$$\theta(0) = \theta_0, \quad \frac{d\theta}{dt}(0) = \theta_1,$$

We introduce two standard approaches for handling nonlinear problems of this kind. One deals with the Taylor series solutions of the problem. The other one uses phase plane analysis and provides qualitative knowledge of the behavior of the solutions. The latter approach is highly geometric. We explore in particular the basis of the latter approach with a view toward the applications.

9.1 The Taylor Series Method

Since (E) is nonlinear, we cannot obtain a general solution in terms of elementary function. However, when t is sufficiently small, we are able to approximate the solution by a Taylor series in the form:

$$\theta(t) = \theta(0) + \theta'(0)\frac{t}{1!} + \theta''(0)\frac{t^2}{2!} + \cdots .$$

To determine the Taylor series for the solution to the initial-value problem $(E)(I)$, we need only determine the values of the derivatives of θ at zero — that is, $\theta(0), \theta'(0), \theta''(0) \cdots$, and so on. The first two terms can be obtained from the given initial conditions and the higher-order terms can be obtained from (E) and its derivatives at $t = 0$. Thus the Taylor series can be completely determined.

9.2   The Phase Plane

The nonlinear equation (E) can be rewritten in the form of a system

$$\frac{dx}{dt} = f(x, y), \quad \frac{dy}{dt} = g(x, y)$$

with $x = \theta, y = d\theta/dt$. Points along a solution of (S) can be viewed as a triple in $\mathbb{R}^3, (t, x(t), y(t))$ — a path traced out in three dimensions, a time coordinate t, and a two-dimensional space coordinate (x, y). The absence of the independent variable t in the right-hand sides of the equations makes another interpretation useful. Solutions may be regarded in the plane as a parametric curve given by $(x(t), y(t))$, with t as the parameter. This curve is called a *trajectory* or an *orbit* and is simply the projection of the triple $(t, x(t), y(t))$ in the three-dimensional space onto the xy- plane of the space variables. The latter is called a *phase plane* of the solution of (S).

A constant solution $x(t) = x_0, y(t) = y_0$ to (S) is called an *equilibrium solution* and its orbit is represented as a single point (x_0, y_0) in the phase plane. Clearly, such points must satisfy the algebraic relations:

(A) $f(x, y) = 0, \quad g(x, y) = 0.$

The points that satisfy (A) are called *equilibrium points* (also *critical points*), and each such point represents an equilibrium solution. The totality of all the orbits of (S) and critical points, graphed in the phase plane, is called the *phase portrait* of (S).

The orbits of the solutions $x = x(t), y = y(t)$ of (S) are also solution curves of the first-order scalar equation:

(DE) $\dfrac{dy}{dx} = \dfrac{g(x, y)}{f(x, y)}$.

Therefore, it is not necessary to find a solution $x(t), y(t)$ of (S) in order to compute its orbit; we need only solve this single first-order scalar differential equation (DE). When this is done, information about how a given orbit depends on the time parameter t is lost; however, this is often not crucial.

In a two-dimensional system of differential equations, there are essentially four types of equilibrium points (equilibria), named for the typical pictures of phase plane trajectories near them: They are classified as *center,spiral, node* and *saddle* as shown here. An equilibrium may be either *stable* or *unstable* according to whether near by trajectories approach or leave that point; the center is stable in the physical sense, but neither attracting nor repelling. Sources (arrows point out) or sinks (arrows point in) are further subdivided into *spiral* and *node*. A *saddle* has some arrows pointing out and some pointing in. The phase portrait shows how x and y interact; it is an extremely important picture. In particular, the phase portrait gives crucial information about the behavior of trajectories near all possible equilibria.

9.3   Linear Systems

In the following, we include here some typical phase portraits for the linear system obtained by using Maple. As is well-known, they are determined by the eigenvalues and eigenvectors of the coefficient matrix of the linear system. We remark that linearization about the equilibrium point of nonlinear system can be achieved by using Taylor's theorem. The local structure of the orbits near the critical point of the nonlinear system can be determined, under fairly broad conditions, by examine the corresponding linearized system. In fact, one can show that *if the real parts of the eigenvalues of the Jacobi matrix (i.e., the coefficient matrix of the linearized system) are not zero, then both nonlinear and their linearized systems have the same qualitative structure in a neighborhood of the critical point.*

```
>   with(DEtools):with(linalg):
```

Warning, new definition for norm

Warning, new definition for trace

```
>   f:=(t,x,y)->a*x+b*y;
```

$$f := (t,\, x,\, y) \rightarrow a\,x + b\,y$$

```
>   g:=(t,x,y)->c*x+d*y;
```

$$g := (t,\, x,\, y) \rightarrow c\,x + d\,y$$

```
>   sys:=diff(x(t),t)=f(t,x,y),
>   diff(y(t),t)=g(t,x,y);
```

$$sys := \frac{\partial}{\partial t}\,\mathrm{x}(t) = a\,x + b\,y,\; \frac{\partial}{\partial t}\,\mathrm{y}(t) = c\,x + d\,y$$

```
>  CrPt:=solve({f(t,x(t),y(t))=0,g(t,x(t),y(t))
>  =0},{x(t),y(t)});
```

$$CrPt := \{x(t) = 0,\ y(t) = 0\}$$

```
>  A0:=GenerateMatrix({f(t,x,y) ,g(t,x,y)},[x,y]);
>
```

$$A0 := \begin{bmatrix} a & b \\ c & d \end{bmatrix}, \begin{bmatrix} 0 \\ 0 \end{bmatrix}$$

$$A := \begin{bmatrix} a & b \\ c & d \end{bmatrix}$$

```
>  inits1:=[[0,0,0],[0,1,0],[0,0,1],[0,-3,0],[0,3,0],[0,0,3],[0,0,2],
>  [0,0,-3],[0,0,3.5],[0,0,2.5],[0,0,-3.5],[0,4,0],[0,-4,0],[0,0,.5],
>  [0,5,0],[0,-5,0],[0,0,-5],[0,Pi,0],[0,-Pi,0]];
>  inits2:=[[0,1,-1],[0,1,1],[0,-1,1],[0,.5,0],[0,0.5,.5],[0,.5,-.5],
>  [0,-.5,.5],[0,0,-.5]];
```

$inits1 := [[0,\ 0,\ 0],\ [0,\ 1,\ 0],\ [0,\ 0,\ 1],\ [0,\ -3,\ 0],\ [0,\ 3,\ 0],\ [0,\ 0,\ 3],\ [0,\ 0,\ 2],\ [0,\ 0,\ -3],$
$[0,\ 0,\ 3.5],\ [0,\ 0,\ 2.5],\ [0,\ 0,\ -3.5],\ [0,\ 4,\ 0],\ [0,\ -4,\ 0],\ [0,\ 0,\ .5],\ [0,\ 5,\ 0],\ [0,\ -5,\ 0],$
$[0,\ 0,\ -5],\ [0,\ \pi,\ 0],\ [0,\ -\pi,\ 0]]$

$inits2 := [$
$[0,\ 1,\ -1],\ [0,\ 1,\ 1],\ [0,\ -1,\ 1],\ [0,\ .5,\ 0],\ [0,\ .5,\ .5],\ [0,\ .5,\ -.5],\ [0,\ -.5,\ .5],\ [0,\ 0,\ -.5]]$

As our first example we consider a matrix which will lead to two real distinct eigenvalues of the same sign.

```
>  A1:=subs({a=-1,b=0,c=0,d=2},op(A));
```

$$A1 := \begin{bmatrix} -1 & 0 \\ 0 & 2 \end{bmatrix}$$

```
>  Eigenvalues(A1);
```

$$2,\ -1$$

```
>  phaseportrait({diff(x(t),t)=-x(t),diff(y(t),t)=2*y(t)},{x(t),y(t)
>  },-10..10,inits2,x=-2..2,y=-2..2, stepsize=0.1,arrows='SLIM');
```

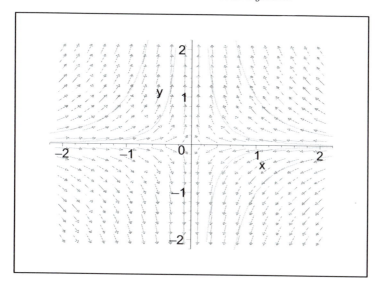

For the second case we choose a matrix which will have distinct, real eigenvalues of opposite sign.

```
>   A2:=subs({a=1,b=0,c=0,d=-2},op(A));
```

$$A2 := \begin{bmatrix} 1 & 0 \\ 0 & -2 \end{bmatrix}$$

$$1, -2$$

```
>   phaseportrait({diff(x(t),t)=x(t),diff(y(t),t)=-2*y(t)},{x(t),y(t)
>   },-10..10,inits2,x=-2..2,y=-2..2, stepsize=0.1,arrows='SLIM');
```

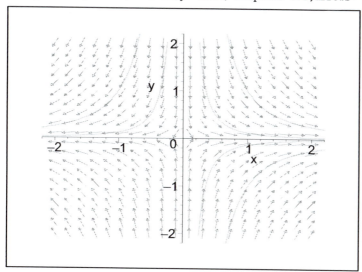

The third case consists of a matrix having complex conjugate eigenvalues.

```
>    A3:=subs({a=1,b=2,c=-2,d=1},op(A));
```

$$A3 := \begin{bmatrix} 1 & 2 \\ -2 & 1 \end{bmatrix}$$

```
>    Eigenvalues(A3);
```

$$1 + 2I, 1 - 2I$$

```
>    phaseportrait({diff(x(t),t)=x(t)+2*y(t),diff(y(t),t)=-2*x(t)+y(t)},
>    {x(t),y(t)},-10..10,inits2,x=-2..2,y=-2..2, stepsize=0.1,
>    arrows='SLIM');
```

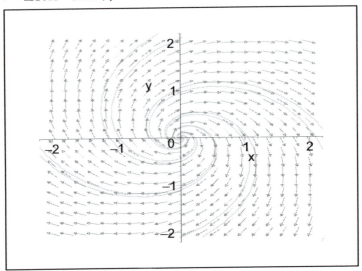

The fourth case is a special case of the preceding one, namely we have a purely imaginary eigenvalue I and its conjugate $-i$.

```
>    A4:=subs({a=0,b=1,c=-1,d=0},op(A));
```

$$A4 := \begin{bmatrix} 0 & 1 \\ -1 & 0 \end{bmatrix}$$

```
>    Eigenvalues(A4);
```

$$I, -I$$

```
>    phaseportrait({diff(x(t),t)=y(t),diff(y(t),t)=-x(t)},{x(t),y(t)
>    },0..10,inits2,x=-2..2,y=-2..2, stepsize=0.1,arrows='SLIM');
```

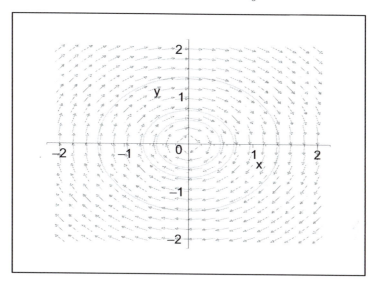

MAPLE is also able do plot the `phaseportrait` of nonlinear autonomous systems. For example, consider the nonlinear problem.

$$\frac{d\,(x(t))}{dt} = -x(t) + \frac{x(t)}{(1+x(t)^2+y(t))^2}$$
$$\frac{d\,(y(t))}{dt} = x(t) + \frac{y(t)}{(1+x(t)^2+y(t)^2)}.$$

```
> phaseportrait(
> {diff(x(t),t)=-x(t)+x(t)/(1+x(t)^2+y(t)^2),diff(y(t),t)=x(t)+y(t)/
> (1+ x(t)^2+y(t)^2)},{x(t),y(t)},-10..10,inits2,x=-2..2,y=-5..5,
> stepsize=0.1, arrows='SLIM');
```

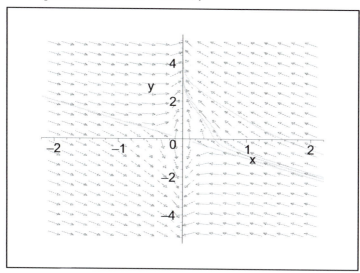

We consider next the non-linear pendulum equation

$$\frac{d^2\theta}{dt^2} + \omega^2 \sin(\theta) = 0,$$

where $\omega^2 := \frac{g}{L}$, g is the gravitational constant and L is the length of the pendulum. If we set

$$x(t) := \theta(t), \text{ and } y(t) := \frac{dx(t)}{dt},$$

we obtain the first-order system

$$\frac{dx(t)}{dt} = y(t)$$
$$\frac{dy(t)}{dt} = -\omega^2 \sin(x(t))$$

We see that MAPLE is able to handel this problem also with a judicious choice of initial data. To this end we chose inits1

```
>   phaseportrait({diff(x(t),t)=y(t),diff(y(t),t)=-sin(
>   x(t))},{x(t),y(t)},-10..10,inits1,x=-10..10,y=-6..6,
>   stepsize=0.01,
>   arrows='SLIM');
```

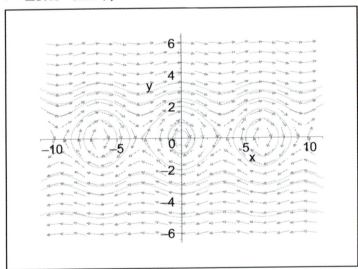

If we put some dampening on the nonlinear pendulum we obtain an equation of the form

$$\frac{d^2\theta}{dt^2} + \gamma\frac{d\theta}{dt} + \omega^2 \sin(\theta) = 0,$$

where γ is a dampening coefficient. By using the same reduction of order, this equation leads to the first-order system

$$\begin{aligned} \frac{d\,x(t)}{dt} &= y(t) \\ \frac{d\,y(t)}{dt} &= -\omega^2 \sin\left(x(t)\right) - cy(t). \end{aligned}$$

We first consider the dampening coefficient to be 0.1, and notice that the solutions resemble the undamped case.

```
> phaseportrait({diff(x(t),t)=y(t),diff(y(t),t)=-sin(
> x(t))-0.1*y(t)},{x(t),y(t)},-5..5,inits1,x=-10..10,y=-6..6,
> stepsize=0.1, arrows='SLIM');
```

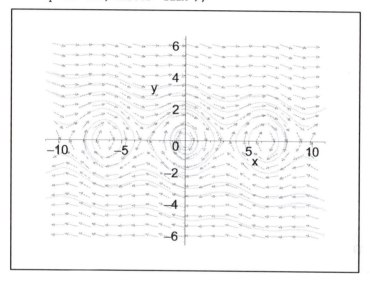

```
> phaseportrait({diff(x(t),t)=y(t),diff(y(t),t)=-sin(
> x(t))-0.2*y(t)},{x(t),y(t)},-5..5,inits1,x=-10..10,y=-6..6,
> stepsize=0.1, arrows='SLIM');
```

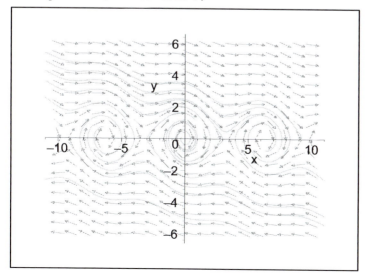

We next increase the dampening to .5 and notice that the nature of the
critical point has changed.

```
>    phaseportrait({diff(x(t),t)=y(t),diff(y(t),t)=-sin(
>    x(t))-0.5*y(t)},{x(t),y(t)},-5..5,inits1,x=-10..10,y=-6..6,
>    stepsize=0.1, arrows='SLIM');
```

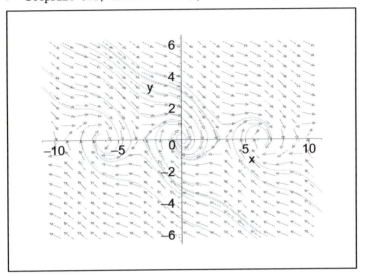

```
>    phaseportrait({diff(x(t),t)=y(t),diff(y(t),t)=-sin(
>    x(t))-y(t)},{x(t),y(t)},-5..5,inits1,x=-10..10,y=-6..6,
>    stepsize=0.1, arrows='SLIM');
```

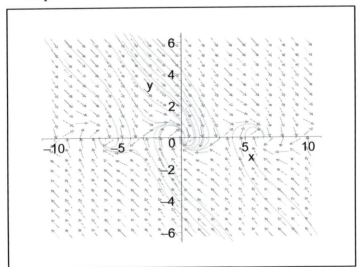

9.4 Useful Maple Commands

We include here a list of Maple commands which may be needed for the computer project below:

`phaseportrait, DEplot2, dsolve`
with options (all in the `DEtools` package)
`phaseplot, rungekuttahf` (all in the `ODEfile`)
`odeplot, implicitplot` (all in the `plots` package).
`map(diff, deq, t)`
`taylor(f(x),x=0,6)`
`convert(",polynom)`

9.5 Computer Lab

To be more specific, we now specify (I) with

$$\theta_0 = \pi/12, \quad \theta_1 = 0.$$

These initial conditions(I) are to be used throughout the following exercises.

(1) Derive the first six terms of the Taylor series about $t = 0$ of the solution to the initial-value problem (E), (I).

(2) Solve the linearized problem defined by (E_0) (I) with

(E_0)
$$\frac{d^2\theta}{dt^2} + \theta = 0.$$

(3) Solve (E)(I) by using `dsolve` with the option 'numeric'.

(4) On the same coordinate axes, graph the approximations found in (1) (2) and the numerical solutions obtained in (3). Comment on the validity of the series solution in (1) and the linearized solution in (2).

(5) Rewrite (E) (I) in the form of the system (S) with initial data, and use `DEplot2` to plot the solution of the system with the initial data (I) in the space (t, x, y).

(6) Find the corresponding nonlinear equation (DE), from which show that the solution of (E) (I) satisfies

(Q)
$$\frac{1}{2}y^2 - \cos x = -\cos\left(\frac{\pi}{12}\right).$$

(7) Solve (DE) in (6) numerically with proper initial data from (Q) with the mesh sizes h = 0.1 and h = 0.01 by using any numerical scheme from Maple.

(8) Plot the following on the same graph: the solution of (Q) in (6) and the numerical solutions in (7). Comment on the validity of the numerical solutions in (7).

(9) Draw a phase portrait for (E$_0$) and plot solution curves $x(t)$ vs t and $y(t)$ vs t beside the phase portrait by using appropriate x and y scales so that one may easily identify their traces on the phase portrait.

(10) Draw a phase portrait for (E) by using (S) in (5) for various initial data.

9.6   Supplementary Maple Programs

9.6.1   Taylor Series Expansion

In this section we show we can use Taylor series methods to solve the nonlinear pendulum problem which we had previously investigated with phaseportrait. First we load DEtools and plots.

```
>   with(DEtools):with(plots):
```

Warning, the name \verb+changecoords+ has been redefined

Next we input the nonlinear pendulum problem.

```
>   ode:=diff(Theta(t),t$2)=-sin(Theta(t));
```

$$ode := \frac{\partial^2}{\partial t^2}\,\Theta(t) = -\sin(\Theta(t))$$

For later purposes of comparison we first solve the initial value problem numerically and store the solution.

```
>   soln:=dsolve({ode,Theta(0)=Pi/12,D(Theta)(0)=0},Theta(t),numeric);
```

$$soln := \textbf{proc}(rkf45\_x) \ldots \textbf{end proc}$$

We make a plot but do not print it!

```
>   g1:=odeplot(soln,[t,Theta(t)],0..10,numpoints=50):
```

In order to obtain a Taylor series solution we expand $\sin(\theta)$ about $\theta = \frac{\pi}{12}$ (Recall the initial condition $\theta(0) = \frac{\pi}{12}$!), and convert this to a polynomial.

```
>   p:=convert(taylor(sin(theta),theta=Pi/12,6),polynom);
```

$$p := \sin(\tfrac{1}{12}\pi) + \cos(\tfrac{1}{12}\pi)\,(\theta - \tfrac{1}{12}\pi) - \tfrac{1}{2}\sin(\tfrac{1}{12}\pi)\,(\theta - \tfrac{1}{12}\pi)^2$$
$$-\tfrac{1}{6}\cos(\tfrac{1}{12}\pi)\,(\theta - \tfrac{1}{12}\pi)^3 + \tfrac{1}{24}\sin(\tfrac{1}{12}\pi)\,(\theta - \tfrac{1}{12}\pi)^4 + \tfrac{1}{120}\cos(\tfrac{1}{12}\pi)\,(\theta - \tfrac{1}{12}\pi)^5$$

As we wish to work with θ we expand this expression into powers of θ.

```
>   p:=expand(%);
```

$$p := \sin(\frac{1}{12}\pi) + \cos(\frac{1}{12}\pi)\theta - \frac{1}{12}\cos(\frac{1}{12}\pi)\pi - \frac{1}{2}\sin(\frac{1}{12}\pi)\theta^2 + \frac{1}{12}\sin(\frac{1}{12}\pi)\theta\pi$$

$$- \frac{1}{288}\sin(\frac{1}{12}\pi)\pi^2 - \frac{1}{6}\cos(\frac{1}{12}\pi)\theta^3 + \frac{1}{24}\cos(\frac{1}{12}\pi)\theta^2\pi - \frac{1}{288}\cos(\frac{1}{12}\pi)\theta\pi^2$$

$$+ \frac{1}{10368}\cos(\frac{1}{12}\pi)\pi^3 + \frac{1}{24}\sin(\frac{1}{12}\pi)\theta^4 - \frac{1}{72}\sin(\frac{1}{12}\pi)\theta^3\pi + \frac{1}{576}\sin(\frac{1}{12}\pi)\theta^2\pi^2$$

$$- \frac{1}{10368}\sin(\frac{1}{12}\pi)\theta\pi^3 + \frac{1}{497664}\sin(\frac{1}{12}\pi)\pi^4 + \frac{1}{120}\cos(\frac{1}{12}\pi)\theta^5$$

$$- \frac{1}{288}\cos(\frac{1}{12}\pi)\theta^4\pi + \frac{1}{1728}\cos(\frac{1}{12}\pi)\theta^3\pi^2 - \frac{1}{20736}\cos(\frac{1}{12}\pi)\theta^2\pi^3$$

$$+ \frac{1}{497664}\cos(\frac{1}{12}\pi)\theta\pi^4 - \frac{1}{29859840}\cos(\frac{1}{12}\pi)\pi^5$$

As we wish θ to be a function of t we rename it as $\varphi(t)$[1]

```
>   p2:=subs(theta=phi(t),p);
```

$$p2 := \sin(\frac{1}{12}\pi) + \cos(\frac{1}{12}\pi)\varphi(t) - \frac{1}{12}\cos(\frac{1}{12}\pi)\pi - \frac{1}{2}\sin(\frac{1}{12}\pi)\varphi(t)^2$$

$$+ \frac{1}{12}\sin(\frac{1}{12}\pi)\varphi(t)\pi - \frac{1}{288}\sin(\frac{1}{12}\pi)\pi^2 - \frac{1}{6}\cos(\frac{1}{12}\pi)\varphi(t)^3 + \frac{1}{24}\cos(\frac{1}{12}\pi)\varphi(t)^2\pi$$

$$- \frac{1}{288}\cos(\frac{1}{12}\pi)\varphi(t)\pi^2 + \frac{1}{10368}\cos(\frac{1}{12}\pi)\pi^3 + \frac{1}{24}\sin(\frac{1}{12}\pi)\varphi(t)^4$$

$$- \frac{1}{72}\sin(\frac{1}{12}\pi)\varphi(t)^3\pi + \frac{1}{576}\sin(\frac{1}{12}\pi)\varphi(t)^2\pi^2 - \frac{1}{10368}\sin(\frac{1}{12}\pi)\varphi(t)\pi^3$$

$$+ \frac{1}{497664}\sin(\frac{1}{12}\pi)\pi^4 + \frac{1}{120}\cos(\frac{1}{12}\pi)\varphi(t)^5 - \frac{1}{288}\cos(\frac{1}{12}\pi)\varphi(t)^4\pi$$

$$+ \frac{1}{1728}\cos(\frac{1}{12}\pi)\varphi(t)^3\pi^2 - \frac{1}{20736}\cos(\frac{1}{12}\pi)\varphi(t)^2\pi^3 + \frac{1}{497664}\cos(\frac{1}{12}\pi)\varphi(t)\pi^4$$

$$- \frac{1}{29859840}\cos(\frac{1}{12}\pi)\pi^5$$

We now rewrite the nonlinear pendulum equation in terms of the unknown $\varphi(t)$.

```
>   ode1:=diff(phi(t),t$2)+p2;
```

---

[1] It is not necessary to change the name of θ to φ, but we avoid confusing the computer if we scroll back to change something .

$$ode1 := (\frac{\partial^2}{\partial t^2}\,\varphi(t)) + \sin(\frac{1}{12}\pi) + \cos(\frac{1}{12}\pi)\,\varphi(t) - \frac{1}{12}\cos(\frac{1}{12}\pi)\,\pi - \frac{1}{2}\sin(\frac{1}{12}\pi)\,\varphi(t)^2$$

$$+ \frac{1}{12}\sin(\frac{1}{12}\pi)\,\varphi(t)\,\pi - \frac{1}{288}\sin(\frac{1}{12}\pi)\,\pi^2 - \frac{1}{6}\cos(\frac{1}{12}\pi)\,\varphi(t)^3 + \frac{1}{24}\cos(\frac{1}{12}\pi)\,\varphi(t)^2\,\pi$$

$$- \frac{1}{288}\cos(\frac{1}{12}\pi)\,\varphi(t)\,\pi^2 + \frac{1}{10368}\cos(\frac{1}{12}\pi)\,\pi^3 + \frac{1}{24}\sin(\frac{1}{12}\pi)\,\varphi(t)^4$$

$$- \frac{1}{72}\sin(\frac{1}{12}\pi)\,\varphi(t)^3\,\pi + \frac{1}{576}\sin(\frac{1}{12}\pi)\,\varphi(t)^2\,\pi^2 - \frac{1}{10368}\sin(\frac{1}{12}\pi)\,\varphi(t)\,\pi^3$$

$$+ \frac{1}{497664}\sin(\frac{1}{12}\pi)\,\pi^4 + \frac{1}{120}\cos(\frac{1}{12}\pi)\,\varphi(t)^5 - \frac{1}{288}\cos(\frac{1}{12}\pi)\,\varphi(t)^4\,\pi$$

$$+ \frac{1}{1728}\cos(\frac{1}{12}\pi)\,\varphi(t)^3\,\pi^2 - \frac{1}{20736}\cos(\frac{1}{12}\pi)\,\varphi(t)^2\,\pi^3 + \frac{1}{497664}\cos(\frac{1}{12}\pi)\,\varphi(t)\,\pi^4$$

$$- \frac{1}{29859840}\cos(\frac{1}{12}\pi)\,\pi^5$$

We now make the type of substitution used in the Chapter on series methods, namely we replace $\varphi(t)$ by a power series with unknown coefficients. The purpose is to use this to determine the Taylor coefficients.

```
>  expr:=simplify(subs(phi(t)=Pi/12+sum(a(m)*t^m,m=2..5),ode1)):
```

We compute the coefficients of the powers of t from 0 until 6, which as we will see, lead to equations for the unknowns a_n.

```
>  for n from 0 to 6 do eq(n):=coeff(expr,t,n) od;
```

$$eq(0) := 2\,a(2) + \sin(\frac{1}{12}\pi)+$$

$$eq(1) := 6\,a(3)$$

$$eq(2) := 12\,a(4) + \cos(\frac{1}{12}\pi)\,a(2)+$$

$$eq(3) := 20\,a(5) + \cos(\frac{1}{12}\pi)\,a(3)$$

$$eq(4) := a(4)\cos(\pi/12) - 1/2\,(a(2))^2\sin(\pi/12)$$
$$eq(5) := a(5)\cos(\pi/12) - a(3)\,a(2)\sin(\pi/12)$$

$$eq(6) := -1/6\,(a(2))^3\cos(\pi/12) + \frac{\left(-120\,a(4)\,a(2) - 60\,(a(3))^2\right)\sin(\pi/12)}{120}$$

We know solve equations 0 through 3 for the unknowns a_2, a_3, a_4, a_5. Computing a few more terms does not appreciably improve the solution.

```
>  solve({eq(0),eq(1),eq(2),eq(3)},{a(2),a(3),a(4),a(5)});
```

$$\{a(2) = -1/2\sin(\pi/12),\, a(3) = 0,\, a(4) = 1/24\sin(\pi/12)\cos(\pi/12),\, a(5) = 0\}$$

These coefficients are then substituted into the series expression for $\varphi(t)$ which gives us the equation to fourth-order accuracy.

$$\theta := \pi/12 - 1/2 \sin(\pi/12)\, t^2 + 1/24 \sin(\pi/12) \cos(\pi/12)\, t^4$$

To obtain numerical values for the purpose of comparison we use `evalf`.

```
>  evalf(%);
```

$$0.2617993878 - 0.1294095226\, t^2 + 0.01041666667\, t^4$$

The Taylor series approximation is then plotted and this is displayed along with the numerical solution we obtained earlier. The series solution agrees with the numerical solution quite well on the interval $[0, 2]$.

```
>  g2:=plot(theta(t),t=0..10,-1.5..1.5,style=POINT,symbol=CIRCLE):
>  display({g1,g2});
```

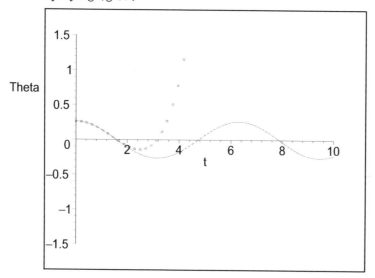

We now make use of MAPLE 's flag `series` and solve the nonlinear pendulum differential this way. The solution is given to fifth-order accuracy and we convert this series to a polynomial. It is exactly the series we had found previously by the power series approach.

```
>  soln_series:=convert(dsolve({ode,Theta(0)=Pi/12,D(Theta)(0)=0
>  },Theta(t),series),polynom);
```

$$soln\_series := \Theta(t) = \pi/12 - 1/2 \sin(\pi/12)\, t^2 + 1/24 \sin\left(\frac{5\pi}{12}\right) \sin(\pi/12)\, t^4$$

9.6.2   The Damped Pendulum

We revisit the dampened pendulum in this section. First we give an alternate plotting procedure using `scene`. The plots are similar to those obtained earlier in this chapter. Then we use the series method to plot the solution for several damping constants. This shows the effect of increasing the damping.

```
>  with(DEtools):
>  inits:=[[x(0)=0,y(0)=1/2],[x(0)=0,y(0)=1],[x(0)=0,y(0)=3/2],[x(0)
>  =0,y(0)=2],[x(0)=0,y(0)=5],[x(0)=0,y(0)=3],[x(0)=0,y(0)=4],[x(0)=0,
>  y(0)=-1]];
```

$inits := [[x(0) = 0, y(0) = \frac{1}{2}], [x(0) = 0, y(0) = 1], [x(0) = 0, y(0) = \frac{3}{2}], [x(0) = 0, y(0) = 2],$

$[x(0) = 0, y(0) = 5], [x(0) = 0, y(0) = 3], [x(0) = 0, y(0) = 4], [x(0) = 0, y(0) = -1]]$

```
>  eq1:=D(x)(t)+y(t)=0;
```

$$eq1 := D(x)(t) + y(t) = 0$$

```
>  eq2:=D(y)(t)-sin(x(t))-0.5*y(t)=0;
```

$$eq2 := D(y)(t) - \sin(x(t)) - .5\,y(t) = 0$$

`scene` specifies the plot to be viewed. For example, `scene=[x,y]` indicates that the plot of x versus y (x horizontal) is to be plotted, with t implicit, while `scene=[t,y]` plots t versus y with t explicit. This option can also be used to change the order in which to plot the variables. There is no default ordering when `vars` is indicated as a set; if `vars` is given as a list, the given ordering will be used.

```
>  DEplot({eq1,eq2},{x(t),y(t)
>  },t=0..15,inits,x=-4..4,y=-4..4,stepsize=0.1,scene=[x(t),y(t)]
>  );
```

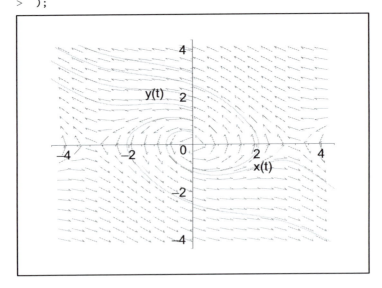

The following `phaseportrait` plots do not effectively show the effect of dampening, hence, we solve the original differential equation for various dampening parameters and `display` these later.

```
>   phaseportrait([eq1,eq2],[x(t),y(t)],t=0..15,[[x(0)=0,y(0)=1/2]],
>   x=-4..4,y=-4..4,stepsize=0.1,scene=[x(t),y(t)]
>   );
```

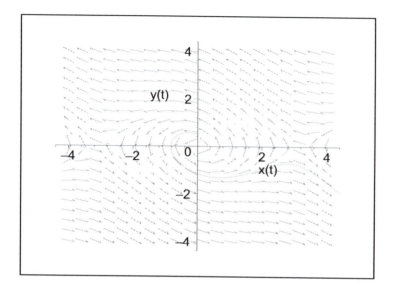

```
>   phaseportrait([eq1,eq2],[x(t),y(t)],t=0..15,inits,x=-4..4,y=-4..4,
>   stepsize=0.1,scene=[x(t),y(t)]
>   );
```

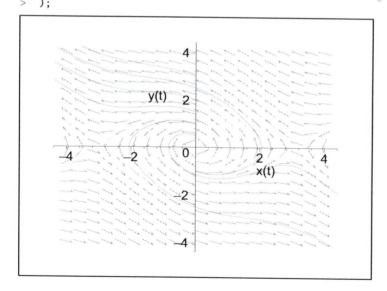

We enter the differential equation for the damped nonlinear pendulum with damping coefficient equal to 0.1.

```
>   ode2:=diff(theta||2(t),t$2)+0.1*diff(theta||2(t),t)+
>   sin(theta||2(t))=0;
```

$$ode2 := \left(\tfrac{\partial^2}{\partial t^2}\,\theta2(t)\right) + 0.1\left(\tfrac{\partial}{\partial t}\,\theta2(t)\right) + \sin(\theta2(t)) = 0$$

Next we use the series flag to obtain a solution.

```
>   soln_series_2:=evalf(convert(dsolve(
>   {ode2,theta||2(0)=Pi/12,D(theta||2)(0)=0
>   },theta||2(t),series),polynom));
```

$soln\_series\_2 := \theta2(t) = 0.2617993878 - 0.1294095226\,t^2 + 0.004313650753\,t^3$
$+\,0.01030882540\,t^4 - 0.0004145098414\,t^5$

Next we consider the same equation with the damping coefficient equal to 0.8.

```
>   ode3:=diff(theta||3(t),t$2)+0.8*diff(theta||3(t),t)+
>   sin(theta||3(t))=0;
```

$$ode3 := \left(\tfrac{\partial^2}{\partial t^2}\,\theta3(t)\right) + 0.8\left(\tfrac{\partial}{\partial t}\,\theta3(t)\right) + \sin(\theta3(t)) = 0$$

```
>   soln_series_3:=evalf(convert(dsolve(
>   {ode3,theta||3(0)=Pi/12,D(theta||3)(0)=0
>   },theta||3(t),series),polynom));
```

$soln\_series\_3 := \theta3(t) = 0.2617993878 - 0.1294095226\,t^2 + 0.03450920600\,t^3$
$+\,0.003514825466\,t^4 - 0.002229038739\,t^5$

Clearly the coefficient are somewhat different, but not appreciably so as can be seen from the graphical comparison here.

```
>   g4:=plot(0.2617993878-0.1294095226*t^2+0.8627301502e-2*t^3
>   +0.9985301595e-2*t^4-0.8160787302e-3*t^5,t=-10..10,-1.5..1.5,
>   style=POINT,symbol=CIRCLE):
>   g5:=plot(0.2617993878-0.1294095226*t^2+0-0.8627301502e-2*t^3
>   +0.9985301595e-2*t^4-0.8160787302e-3*t^5,t=-10..10,-1.5..1.5,
>   style=LINE,symbol=CROSS):
>   display({g4,g5});
```

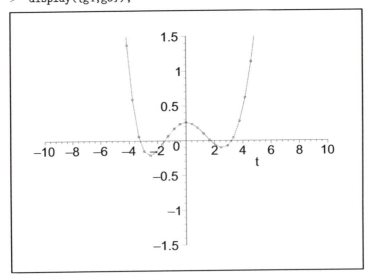

We now seek numerical solutions of these equations.

```
>   soln_2:=dsolve({ode2,theta||2(0)=Pi/12,D(theta||2)(0)=0
>   },theta||2(t),numeric);
```

$$soln\_2 := \mathbf{proc}(rkf45\_x) \dots \mathbf{end\ proc}$$

```
>   soln_3:=dsolve({ode3,theta||3(0)=Pi/12,D(theta||3)(0)=0
>   },theta||3(t),numeric);
```

$$soln\_3 := \mathbf{proc}(rkf45\_x) \dots \mathbf{end\ proc}$$

```
>   g6:=odeplot(soln_2,[t,theta||2(t)],-1..10,numpoints=50):
```

```
>   g7:=odeplot(soln_3,[t,theta||3(t)],-1..10,numpoints=50):
```

The display below shows the undamped solution as periodic, the slightly damped as somewhat smaller by the next cycle, and the heavily damped as almost annihilated.

```
>   display({g1,g6,g7});
```

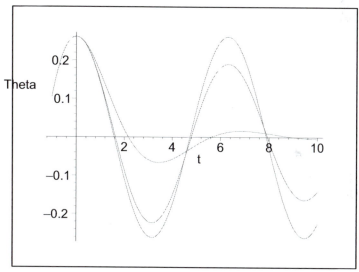

Chapter 10

Integral Transforms

10.1 The Laplace Transform of Elementary Functions

The LAPLACE transform is useful for solving ordinary and partial differential equations. The function $f(s)$ defined by the improper integral as

$$\hat{f}(s) = \int_0^\infty e^{-sx} f(x)\, dx\,,$$

is called the Laplace transform of f and is customarily denoted by $\hat{f} = \mathcal{L}(f)$. The function f is usually taken to be a piecewise continuous function on $[0, \infty)$. A simple condition on $\hat{f}$ which guarantees the existence of the integral in (10.1) (besides the piecewise continuity) is the existence of constants γ and $c > 0$ such that

$$|f(x)| < ce^{\gamma x}\,, \quad \text{for } 0 \le x < \infty\,.$$

Extensive tables of Laplace transforms for various functions f exist. We state the following.

Theorem 10.1. *Suppose* $f \in C\,[0, \infty)$ *and is such that*

$$|f(x)| < ce^{\gamma x}, \quad 0 \le x < \infty\,.$$

If $\hat{f}$ *is the Laplace transform of* f *then*

$$\hat{f}^{(k)}(s) = (-1)^k \int_0^\infty e^{-sx} x^k f(x)\, dx\,,$$

where k *is an arbitrary nonnegative integer.*

The theorem states the Laplace transform $\hat{f}$ possesses derivatives of all orders for $s \ge s_0 > \gamma$ and, conversely, provides a way to compute the Laplace transforms of

$$(-1)^k x^k f(x)\,,$$

given the Laplace transform of $f(x)$. Thus it is easy to verify

$$\frac{1}{s} = \int_0^\infty e^{-sx} \cdot 1 \cdot dx,$$

so the Laplace transform of x^k is $k!/s^{k+1}$, $k = 1, 2, \cdots$. We list some properties of Laplace transforms here. In each case we assume $f(x)$ is piecewise continuous and exponentially bounded with exponent γ. If $\hat{f}(s) = \mathcal{L}(f(x))$, then

$$\mathcal{L}(e^{ax}f(x)) = \int_0^\infty e^{-(s-a)x} f(x)\, dx = \hat{f}(s-a).$$

Note $s - a$ must be large than γ, so $s > \gamma + a$ for the result to be valid. Another result is

$$\mathcal{L}(f(ax)) = \frac{1}{a}\hat{f}(s/a), \quad a > 0;$$

indeed, under the change of variable $ax = \xi$, $\mathcal{L}(f(ax))$ is transformed to

$$\int_0^\infty e^{-s\xi/a} f(\xi)\frac{d\xi}{a} = \frac{1}{a}\hat{f}(s/a).$$

As a final property of Laplace transforms we note that if $\mathcal{L}(f(x)) = \hat{f}(s)$, then

$$\mathcal{L}\left(\int_0^x \hat{f}(y)\, dy\right) = \frac{\hat{f}(s)}{s}.$$

These calculations are valid if f is piecewise continuous and exponentially bounded. We now use MAPLE to find the Laplace transforms of some of the elementary functions of the calculus.

```
>  laplace(exp(alpha*t),t,s);
```

$$\frac{1}{s - \alpha}$$

```
>  laplace(cos(t),t,s);
```

$$\frac{s}{s^2 + 1}$$

```
>  laplace(sin(t),t,s);
```

$$\frac{1}{s^2 + 1}$$

```
>  laplace(t^n,t,s);
```

$$\frac{\Gamma(n+1)}{s^{(n+1)}}$$

```
>  laplace(diff(y(t),t),t,s);
```

$$\text{laplace}(y(t),\ t,\ s)\ s - y(0)$$

We may use MAPLE to compute the Laplace transforms of some typical functions and thereby verify our table of transforms which we have found by hand computation. With this end in mind we consider the trigonometric functions $\sin(at)$, and $\cos(at)$.

> `laplace(sin(a*t),t,s);`

$$\frac{a}{a^2 + s^2}$$

> `laplace(cos(a*t),t,s);`

$$\frac{s}{a^2 + s^2}$$

We next consider the transforms of the exponential functions $\exp(a + ib)t$.

> `laplace(exp((alpha+ I*beta)*t),t,s);`

$$\frac{1}{s - \alpha - I\,\beta}$$

> `laplace(exp((alpha+ I*beta)*t) +exp(-(alpha+`
> `I*beta)*t),t,s);`

$$2\,\frac{s}{s^2 - (\alpha + i\beta)^2}$$

> `simplify(%);`

$$-2\,\frac{s}{(\alpha + i\beta + s)(\alpha + i\beta - s)}$$

Use the above formulas to obtain the Laplace transforms for $\exp at\,\sin(bt)$ and $\exp at\,\cos(bt)$. We consider next the differential equation

> `laplace(diff(y(t),t$2),t,s);`

$$s^2\,laplace\,(y(t),t,s) - D(y)(0) - sy(0)$$

> `laplace(diff(y(t),t$3),t,s);`

$$s^3\,laplace\,(y(t),t,s) - \left(D^{(2)}\right)(y)(0) - sD(y)(0) - s^2 y(0)$$

Next we find the Laplace transforms of the hyperbolic functions $\sinh(\alpha t)$ and $\cosh(\alpha t)$.

> `laplace(sinh(alpha*t),t,s);`

$$\frac{\alpha}{-\alpha^2 + s^2}$$

> `laplace(cosh(alpha*t),t,s);`

$$\frac{s}{-\alpha^2 + s^2}$$

Next we find the Laplace transforms of the hyperbolic functions $\exp at \sinh(bt)$ and $\exp at \cosh(bt)$.

> `laplace(exp(a*t)*sin(b*t),t,s);`

$$\frac{b}{(s-a)^2 + b^2}$$

> `laplace(exp(a*t)*cos(b*t),t,s);`

$$\frac{s-a}{(s-a)^2 + b^2}$$

> `laplace(exp(a*t)*sinh(b*t),t,s);`

$$\frac{b}{(s-a)^2 - b^2}$$

> `laplace(exp(a*t)*cosh(b*t),t,s);`

$$\frac{s-a}{(s-a)^2 - b^2}$$

Here we find the Laplace transform of a power of t times an exponential function of t.

> `laplace(t^n*exp(a*t),t,s);`

$$\frac{\Gamma(n+1)}{(s-a)^{(n+1)}}$$

The Heaviside function is defined as

$$\text{Heaviside}(t) := \begin{cases} 0 & \text{when } x < 0 \\ 1 & \text{when } x \geq 0. \end{cases}$$

We find its Laplace transform below.

> `laplace(Heaviside(t),t,s);`

$$\frac{1}{s}$$

The Dirac delta function is not in the usual sense of the word a function. It may be defined as a **generalized function** or as a **functional**. We designate the Dirac function as $\delta(t-t_0)$, and it has the property that if $f(t)$ is continuous on $(-\inf, \inf)$ then

$$\int_{-\infty}^{\infty} \delta(t-t_0)f(t)\, dt = f(t_0).$$

The delta function may be defined as a limit of a sequence of functions, for example one such sequence is

$$\varphi_n(x)) := \begin{cases} n & \text{when } |t - t_0| \le 1/n \\ 0 & \text{when } |t - t_0| > 1/n. \end{cases}$$

MAPLE knows the delta function is the derivative of the Heaviside function. Let us check this out by recognizing if the delta function is the derivative of the Heaviside function, then the Laplace transform of the delta function should be s times the Laplace transform of the Heaviside function. Why?

```
>   laplace(Dirac(t),t,s);
```

$$1$$

```
>   laplace(Heaviside(t -1),t,s);
```

$$\frac{e^{(-s)}}{s}$$

```
>   laplace(Heaviside(t -25),t,s);
```

$$\frac{e^{(-25\,s)}}{s}$$

```
>   laplace(Heaviside(t -c),t,s);
```

$$\text{laplace}(\text{Heaviside}(t - c),\, t,\, s)$$

The Bessel functions turn up in many problems involving cylindrical co-ordinates. Hence these are very useful functions when solving the differential equations for a vibrating circular membrane. MAPLE knows about these functions too. The Bessel function of the first kind and zeroth order is denoted by $J_0(t)$

$$t^2 \frac{d^2 y(t)}{dt^2} + t \frac{dy(t)}{dt} + t^2 y(t) = 0.$$

A series representation for this function is given by

$$J_0(t) := 1 + \sum_{n=1}^{\infty} \frac{(-1)^n t^{2n}}{2^{2n}(n!)^2}.$$

```
>   laplace(BesselJ(0,t),t,s);
```

$$\frac{1}{\sqrt{s^2 + 1}}$$

An interesting result concerns the Laplace transform of $J_0(\sqrt{t})$.

```
>   laplace(BesselJ(0, sqrt(t)),t,s);
```

$$\frac{1}{s} e^{-1/4\,s^{-1}}$$

```
>   laplace(-t*f(t),t,s);
```

$$\frac{\partial}{\partial s} \text{laplace}(f(t), \, t, \, s)$$

If we do not specify that n is a specific integer MAPLE cannot evaluate the following

> `laplace((-t)^n*f(t),t,s);`

$$\text{laplace}((-t)^n \, f(t), \, t, \, s)$$

However, it is clear what the rule is from the following example. The student should check this with other integers.

> `laplace((-t)^5*f(t),t,s);`

$$\frac{\partial^5}{\partial s^5} \text{laplace}(f(t), \, t, \, s)$$

We list the results of the above computer investigation as a table.

formula	$\widehat{f}(s)$	$f(x)$
1	$1/s$	1
2	$1/(s-a)$	e^{ax}
3	$1/s^{n+1}$	$x^n/n! \quad n = 0, 1, 2, \cdots$
4	$1/(s-a)^{n+1}$	$e^{ax}x^n/n! \quad n = 0, 1, 2, \cdots$
5	$1/(s^2 + a^2)$	$(\sin ax)/a$
6	$s/(s^2 + a^2)$	$\cos ax$
7	$1/(s^2 - a^2)$	$(\sinh ax)/a$
8	$s/(s^2 - a^2)$	$\cosh ax$
9	$1/((s-b)^2 + a^2)$	$e^{bx}(\sin ax)/a$
10	$s/((x-b)^2 + a^2)$	$e^{bx}(\cos ax + (b/a)\sin ax)$
11	$1/((s-b)^2 - a^2)$	$e^{bx}(\sinh ax)/a$
12	$s/((s-b)^2 - a^2)$	$e^{bx}(\cosh ax + (b/a)\sinh ax)$

10.2   Solving Differential Equations with the Laplace Transform

In this session we use MAPLE to solve ordinary differential equations using the Laplace transform method. The Laplace transform of a piecewise continuous function $f(t)$ defined on the half -axis $[0, \inf)$ is defined to be

$$\widehat{f}(s) \; := \; \mathcal{L}(f)(s) \; := \; \int_0^\infty \exp{-st} f(t) \, dt.$$

The Laplace transform of a derivative of a piecewise differentiable function, having continuous derivatives of order at least n is given by

$$\mathcal{L}(f^{(n)}(s) = s^n f(\hat{s}) - s^{n-1} f(o) - \cdots - f^{(n-1)}(0).$$

This formula allows us to transform an ordinary differential equation into an algebraic equation for the transformed function $\hat{f}(s)$, including the initial data as given in the successive derivatives of $f(t)$ up to oder $n-1$. We show how this method works by applying the Laplace transform to several differential equations. The first example is the homogeneous equation We now consider how we can solve differential equations using the Laplace transform with MAPLE. For e.xpository purposes let us first consider the differential equation

$$\frac{d^2 y(t)}{dt^2} + \alpha^2 y(t) = 0.$$

```
>   ode1:=diff(y(t),t$2) + alpha^2 *y(t) =0:
>   with(inttrans);
```

[*addtable, fourier, fouriercos, fouriersin, hankel, hilbert, invfourier, invhilbert, invlaplace, invmellin, laplace, mellin, savetable*]

For convenience let us rename the Laplace transform of the solution $Y(s)$.

```
>   Y(s):=solve(laplace(ode1,t,s),
>   laplace(y(t),t,s));
```

$$Y(s) := -\frac{-s\,y(0) - D(y)(0)}{\alpha^2 + s^2}$$

As we will want to invert the Laplace transform, i.e. compute $(L^{-1}Y)(t)$ we need to read in the collection of Laplace transforms and their inverses.

```
>   readlib(laplace):
>   invlaplace(Y(s),s,t);
```

$$y(0)\cos(\alpha\,t) + \frac{D(y)(0)\sin(\alpha\,t)}{\alpha}$$

We first consider a differential equation with a non-homogeneous term on the right.

```
>   ode1:=diff(y(x),x$2)+y(x)=sin(x);
```

$$ode1 := (\tfrac{\partial^2}{\partial x^2}\,y(x)) + y(x) = \sin(x)$$

Differential Equations: A Maple^{TM} Supplement, Second Edition

Because the differential has constant coefficients in the homogeneous part by applying the Laplace transform we obtain an algebraic equation which we may solve for the transformed unknown as shown below.

> laplace(ode1,x,s);

$$s^2 \, laplace\,(y\,(x)\,,x,s) - D\,(y)\,(0) - sy\,(0) + laplace\,(y\,(x)\,,x,s) = \left(s^2 + 1\right)^{-1}$$

> solve(%,laplace(y(x),x,s));

$$\frac{y\,(0)\,s^3 + D\,(y)\,(0)\,s^2 + sy\,(0) + D\,(y)\,(0) + 1}{\left(s^2 + 1\right)^2}$$

At this point we are able to solve for the general solution of the differential equation by computing the inverse Laplace transform. First we read in the library of Laplace transforms.

> readlib(laplace);

proc(*expr*::{*set, algebraic, equation*}, *t*::*name, s*::*algebraic*) ... **end proc**

> invlaplace((y(0)*s^3+s*y(0)+D(y)(0)*s^2+D(y)(0)+1)/
> (s^4+2*s^2+1),s,t);

$$- 1/2\,\cos{(t)}\,(-2\,y\,(0) + t) + 1/2\,\sin{(t)}\,(1 + 2\,D\,(y))$$
$$-1/2\,\cos{(t)}\,(-2\,y\,(0) + t) + 1/2\,\sin{(t)}\,(1 + 2\,D\,(y)\,(0))\,(0))$$

So we have the general solution which may be made definite by substituting in the initial values. We try next a nonhomogeneous equation where the right hand side does not permit the use of undetermined coefficients.

> ode2:=diff(y(x),x$2)-2*y(x)=x^2*exp(x^2);

$$ode2 := (\tfrac{\partial^2}{\partial x^2}\,y(x)) - 2\,y(x) = x^2\,e^{(x^2)}$$

> laplace(ode2,x,s);

$$s^2\,laplace\,(y\,(x)\,,x,s) - D\,(y)\,(0) - sy\,(0) - 2\,laplace\,(y\,(x)\,,x,s) = laplace\left(x^2 e^{x^2},x,s\right)$$

> solve(%,laplace(y(x),x,s));

$$\frac{s\,y(0) + D(y)(0) + (\tfrac{\partial^2}{\partial s^2}\,laplace(e^{(x^2)},\,x,\,s))}{s^2 - 2}$$

Again we have obtained an algebraic equation for $Y(s)$ which we invert directly as:

> invlaplace((s*y(0)+D(y)(0)+diff(laplace(exp(x^2),x,s),
> '$'(s,2)))/(s^2-2),s,x);

$$\frac{1}{4}\,e^{(x^2)} + (-\frac{1}{8} - \frac{1}{4}\,D(y)(0)\,\sqrt{2} + \frac{1}{2}\,y(0))\,e^{(-\sqrt{2}\,x)} + (\frac{1}{4}\,D(y)(0)\,\sqrt{2} + \frac{1}{2}\,y(0) - \frac{1}{8})\,e^{(\sqrt{2}\,x)}$$

does not have constant coefficients in the homogeneous part and this makes the problem much more difficult for the Laplace transform method.

> ode3:=t*diff(y(t),t$2)+diff(y(t),t)+t*y(t)=0;

$$ode3 := t\,(\tfrac{\partial^2}{\partial t^2}\,y(t)) + (\tfrac{\partial}{\partial t}\,y(t)) + t\,y(t) = 0$$

> laplace(ode3,t,s);

$$-s \ laplace \left(y\left(t\right),t,s\right) - s^2 \frac{\partial}{\partial s} laplace \left(y\left(t\right),t,s\right) - \frac{\partial}{\partial s} laplace \left(y\left(t\right),t,s\right) = 0$$

Again we replace the Laplace transformed unknown by $Y(s)$ and this time obtain a differential equation for $Y(s)$. This may not be an advantage, but in this instance it works.

```
>  subs(laplace(y(t),t,s)=Y(s),%);
```

$$-sY\left(s\right) - s^2 \frac{d}{ds} Y\left(s\right) - \frac{d}{ds} Y\left(s\right) = 0$$

We solve the differential equation for $Y(s)$ with `dsolve` to obtain:

```
>  dsolve(%,Y(s));
```

$$Y(s) = -\frac{C1}{\sqrt{s^2+1}}$$

The next equation The inverse Laplace transform library can solve this problem readily, however.[1]

```
>  invlaplace(_C1/(s^2+1)^(1/2),s,t);
```

$$\_C1 \ BesselJ(0,\ t)$$

This result should have been anticipated as the differential equation was Bessel's differential equation of oder 0.

To see how easily the Laplace transform method can become flaunted we need simply consider the confluent, hypergeometric equation [5] equation 2.113. This equation is hardly more complicated that the previous equation except for the two constants a and b and is an equation simply solved by the

```
>  ode4:=t*diff(y(t),t$2)+(b-t)*diff(y(t),t)-a*y(t)=0;
```

$$ode4 := t\left(\frac{\partial^2}{\partial t^2} y(t)\right) + (b-t)\left(\frac{\partial}{\partial t} y(t)\right) - a\,y(t) = 0$$

```
>  laplace(ode4,t,s);
```

$$laplace\left(y\left(t\right),t,s\right) s\, b - s^2 \frac{\partial}{\partial s} laplace\left(y\left(t\right),t,s\right) - 2\, laplace\left(y\left(t\right),t,s\right) s$$

$$-a\, laplace\left(y\left(t\right),t,s\right)+s\frac{\partial}{\partial s} laplace\left(y\left(t\right),t,s\right)-y\left(0\right)b+laplace\left(y\left(t\right),t,s\right)+y\left(0\right) = 0$$

Again we set $Y(s)$ equal to the transformed unknown, namely

```
>  subs(laplace(y(t),t,s)=Y(s),%);
```

$$-s\, Y(s) + y(0) - s\left(Y(s) + s\left(\tfrac{\partial}{\partial s} Y(s)\right)\right) + b\left(s\, Y(s) - y(0)\right) + Y(s) + s\left(\tfrac{\partial}{\partial s} Y(s)\right) - a\, Y(s) = 0$$

```
>  simplify(%);
```

$$Y\left(s\right) sb - s^2 \frac{d}{ds} Y\left(s\right) - 2\,Y\left(s\right) s - a Y\left(s\right) + s\frac{d}{ds} Y\left(s\right) - y\left(0\right)b + Y\left(s\right) + y\left(0\right) = 0$$

The solution of the equation for $Y(s)$ is quite complicated, however, and may be seen to be expressible in terms of hypergeometric functions.

---

[1] If MAPLE does not know the inversion we may frequently directly do the inversion ourselves by means of complex integration. This is, however, beyond the scope of the present text.

```
>    dsolve(%,Y(s));
```

$Y(s) = s^{(-1+a)} (s-1)^{(-1+b-a)} \_C1 + s^{(-1+a)} (s-1)^{(-1+b-a)} (-1)^{(a-b)} y(0) s^{(1-a)}$
hypergeom$([-a+b, 1-a], [2-a], s)(-1+b)/(-1+a)$

However, inverting this expression for $Y(s)$ is quite formidable and we stop here. The hypergeometric equation [10] Chapter 2, is easier to solve using another method, for example the Frobenius series method. Indeed from dsolve we obtain a solution in terms of the Kummer functions [5, 11]

```
>    dsolve(ode4,y(t));
```

$$y(t) = \_C1 \, M(a, b, t) + \_C2 \, U(a, b, t)$$

As a further example we consider another equation with a variable coefficient on the right hand side. It turns out that even though we obtain a differential equation for $Y(s)$ it has a solution that we may invert.

```
>    ode5:=diff(y(x),x$2)+x*diff(y(x),x)+y(x)=0;
```

$$ode5 := \left(\tfrac{\partial^2}{\partial x^2} y(x)\right) + x \left(\tfrac{\partial}{\partial x} y(x)\right) + y(x) = 0$$

```
>    laplace(ode5,x,s);
```

$$s^2 \, laplace\,(y(x), x, s) - D(y)(0) - sy(0) - s\frac{\partial}{\partial s} laplace\,(y(x), x, s) = 0$$

```
>    subs(laplace(y(x),x,s)=Y(s),%);
```

$$s^2 Y(s) - D(y)(0) - sy(0) - s\frac{d}{ds}Y(s) = 0$$

```
>    dsolve(%,Y(s));
```

$$s^2 Y(s) - D(y)(0) - sy(0) - s\frac{d}{ds}Y(s) = 0$$

```
>    invlaplace((-1/2*y(0)*sqrt(Pi)*sqrt(2)*erf(1/2*sqrt(2)*s)+1/2*D(y)
>    (0)*Ei(1,1/2*s^2)+_C1)*exp(1/2*s^2),s,x);-1/2*y(0)*sqrt(Pi)*sqrt(2)
>    *invlaplace(exp(1/2*s^2),s,x)+y(0)*exp(-1/2*x^2)+1/2*D(y)(0)*
>    invlaplace (exp(1/2*s^2)*Ei(1,1/2*s^2),s,x)+_C1*invlaplace
>    (exp(1/2*s^2),s,x);
```

$-\tfrac{1}{2} y(0) \sqrt{\pi} \sqrt{2}\, \text{invlaplace}(e^{(1/2 s^2)}, s, x) + y(0) e^{(-1/2 x^2)}$

$+\tfrac{1}{2} D(y)(0)\, \text{invlaplace}(e^{(1/2 s^2)}\, \text{Ei}(1, \tfrac{1}{2} s^2), s, x) + \_C1\, \text{invlaplace}(e^{(1/2 s^2)}, s, x)$

$-\tfrac{1}{2} y(0) \sqrt{\pi} \sqrt{2}\, \text{invlaplace}(e^{(1/2 s^2)}, s, x) + y(0) e^{(-1/2 x^2)}$

$+\tfrac{1}{2} D(y)(0)\, \text{invlaplace}(e^{(1/2 s^2)}\, \text{Ei}(1, \tfrac{1}{2} s^2), s, x) + \_C1\, \text{invlaplace}(e^{(1/2 s^2)}, s, x)$

In the above, Ei denotes the exponential integral defined by $Ei(x) = \int_x^\infty \frac{e^{-t}}{t} dt$, while for non-negative integer n,

$$Ei(n, x) := \int_1^\infty \frac{e^{-xt}}{t^n} dt.$$

We check that this is really the correct answer by using dsolve on the differential equation ode5.

> dsolve(ode5,y(x));

$$y(x) = \frac{\operatorname{erf}(\frac{1}{2} I \sqrt{2} x) \_ C1}{e^{(1/2\,x^2)}} + \frac{\_C2}{e^{(1/2\,x^2)}}$$

Here erf is the error function defined by $erf(x) := \frac{1}{\sqrt{\pi}} \int_0^x e^{-t^2} dt$. Having seen where we get difficulties with the Laplace transform method, let us consider some equations where it works quite easily. To this end we consider the non-homogeneous equation

$$\frac{d^2 y(t)}{dt^2} + 4y(t) = t^2 + 5.$$

This equation can be solved easily by the method of undetermined coefficients.

> deq6:=diff(y(t),t$2) + 4*y(t)- t^2+5;

$$deq6 := (\frac{\partial^2}{\partial t^2} y(t)) + 4\,y(t) - t^2 + 5$$

> laplace(deq6,t,s);

$s^2\, laplace\,(y\,(t),t,s) - D\,(y)\,(0) - sy\,(0) + 4\, laplace\,(y\,(t),t,s) - 2\,s^{-3} + 5\,s^{-1}$

As before rename $\hat{y}(s)$ by $Y(s)$.

> subs(laplace(y(t),t,s)=Y(s),%);

$$s^2 Y\,(s) - D\,(y)\,(0) - sy\,(0) + 4\,Y\,(s) - 2\,s^{-3} + 5\,s^{-1}$$

> solve(%,Y(s));

$$\frac{s^4\, y(0) + D(y)(0)\, s^3 - 5\,s^2 + 2}{s^3\,(s^2 + 4)}$$

Having solved algebraically for the Laplace transformed function $\hat{f}(s)$, we proceed to to use MAPLE to find the **inverse Laplace transform.**

> invlaplace(%,s,t);

$$-\frac{11}{8} + 1/2\,D\,(y)\,(0)\sin\,(2\,t) + 1/4\,t^2 + 1/8\,\cos\,(2\,t)\,(8\,y\,(0) + 11)$$

The situation where the Laplace transform is really useful in the solving of ordinary differential equations is where the nonhomogeneous term is discontinuous and the coefficients on the left hand side are constants. With this illustration in mind we consider a differential equation with a delta function on the right. As you recall from earlier MAPLE calls this function **Dirac**. Let us try to use the Laplace transform and MAPLE to solve a non-homogeneous equation.

$$\frac{d^2 y(t)}{dt^2} + \frac{dy(t)}{dt} + y(t) = \delta(x - 2).$$

> with(inttrans):
> ode7:=diff(y(t),t$2) +diff(y(t),t)+y(t)=Dirac(t-2);

$$ode7 := (\tfrac{\partial^2}{\partial t^2}\, y(t)) + (\tfrac{\partial}{\partial t}\, y(t)) + y(t) = Dirac(t-2)$$

\> laplace(ode7,t,s);

$$s^2\, laplace\,(y\,(t)\,,t,s) - D\,(y)\,(0) - sy\,(0) + s\,laplace\,(y\,(t)\,,t,s) - y\,(0)$$
$$+ laplace\,(y\,(t)\,,t,s)\; =\; \mathrm{e}^{-2\,s}$$

\> subs(laplace(y(t),t,s)=Y(s),%);

$$s^2 Y\,(s) - D\,(y)\,(0) - sy\,(0) + sY\,(s) - y\,(0) + Y\,(s) = \mathrm{e}^{-2\,s}$$

\> solve(%,Y(s));

$$\frac{s\,y(0) + y(0) + D(y)(0) + e^{(-2\,s)}}{s^2 + s + 1}$$

\> invlaplace((s*y(0)+D(y)(0)+y(0)+exp(-2*s))/(s^2+s+1),s,t);

$$y(0)\, e^{(-1/2\,t)} \cos(\tfrac{1}{2}\,\sqrt{3}\,t) + \tfrac{2}{3}\,\mathrm{Heaviside}(t-2)\,\sqrt{3}\,e^{(-1/2\,t+1)}\sin(\tfrac{1}{2}\,\sqrt{3}\,(t-2))$$
$$+\tfrac{1}{3}\,\sqrt{3}\,(y(0) + 2\,D(y)(0))\sin(\tfrac{1}{2}\,\sqrt{3}\,t)\,e^{(-1/2\,t)}$$

As another example we consider the right hand side to be a Heaviside function.

\> ode8:=diff(y(t),t$2) +2*diff(y(t),t)+y(t)=Heaviside(t-2);

$$ode8 := (\tfrac{\partial^2}{\partial t^2}\, y(t)) + 2\,(\tfrac{\partial}{\partial t}\, y(t)) + y(t) = \mathrm{Heaviside}(t-2)$$

\> laplace(ode8,t,s);

$$s^2\, laplace\,(y\,(t)\,,t,s) - D\,(y)\,(0) - sy\,(0) + 2\,s\,laplace\,(y\,(t)\,,t,s) - 2\,y\,(0)$$
$$+ laplace\,(y\,(t)\,,t,s)\; =\; \frac{\mathrm{e}^{-2\,s}}{s}$$

\> subs(laplace(y(t),t,s)=Y(s),%);

$$s^2 Y\,(s) - D\,(y)\,(0) - sy\,(0) + 2\,sY\,(s) - 2\,y\,(0) + Y\,(s) = \frac{\mathrm{e}^{-2\,s}}{s}$$

\> solve(%,Y(s));

$$\frac{s^2\,y(0) + 2\,s\,y(0) + D(y)(0)\,s + e^{(-2\,s)}}{s\,(s^2 + 2\,s + 1)}$$

\> invlaplace(%,s,t);

$$\mathrm{e}^{-t}\,(y\,(0)\,t + D\,(y)\,(0)\,t + y\,(0)) + \left(1 - \mathrm{e}^{2-t}\,(t-1)\right)\mathrm{Heaviside}\,(t-2)$$

We try one more example of a discontinuous right hand side. The reader should experiment with examples of this type.

\> ode9:=diff(y(t),t$2)
\> +2*diff(y(t),t)+y(t)=t*Heaviside(t-2)*sin(t-2);

$$ode9 := (\tfrac{\partial^2}{\partial t^2}\, y(t)) + 2\,(\tfrac{\partial}{\partial t}\, y(t)) + y(t) = t\,\mathrm{Heaviside}(t-2)\sin(t-2)$$

\> laplace(ode9,t,s);

$$s\,(s\,laplace(y(t),\,t,\,s) - y(0)) - D(y)(0) + 2\,s\,laplace(y(t),\,t,\,s) - 2\,y(0) + laplace(y(t),\,t,\,s)$$
$$= 2\,\frac{e^{(-2\,s)}}{s^2+1} + \frac{2\,e^{(-2\,s)}\,s}{(s^2+1)^2}$$

```
>  solve(%,laplace(y(t),t,s));
>  ;
```

$$e^{-t}\left(y\left(0\right)t+\mathrm{D}\left(y\right)\left(0\right)t+y\left(0\right)\right)+1/2\left(\sin\left(t-2\right)+\left(-\cos\left(t-2\right)+e^{2-t}\right)\left(t-1\right)\right)$$
$$Heaviside\left(t-2\right)$$

10.3 Fourier Transforms

In this section we shall introduce several other transforms that MAPLE can use. These are the Fourier (exponential) transform , the Fourier sine transform, and the Fourier cosine transform. These are defined as

$$F(\xi) := \mathcal{F}(f(x), x \to \xi) := \int_{-\infty}^{\infty} f(x)\exp i\xi x\,dx,$$

$$F_c(\xi) := \mathcal{F}_c(f(x), x \to \xi) := \sqrt{\frac{2}{\pi}} \int_0^{\infty} f(x)\cos \xi x\,dx,$$

and

$$F_s(\xi) := \mathcal{F}_s(f(x), x \to \xi) := \sqrt{\frac{2}{\pi}} \int_0^{\infty} f(x)\sin \xi x\,dx,$$

respectively. The inverse of the Fourier transform is

$$f(x) := \mathcal{F}^{-1}(F(\xi), \xi \to x) := \frac{1}{2\pi} \int_{-\infty}^{\infty} f(\xi)\exp \xi x\,d\xi;$$

whereas, the fourier sine and cosine transforms are their own inverses. In order to use integral transform package in MAPLE we must first load inttransforms. Notice it contains other integral transforms, such as the HILBERT and MELLIN transforms.

```
>  with(inttrans);
```

$[addtable, fourier, fouriercos, fouriersin, hankel, hilbert, invfourier, invhilbert, invlaplace, laplace, mellin]$

Let us try the Fourier sine transform on the function $f(t) := t$. We obtain

```
>  fouriersin(t,t,omega);
```

$$-1/2\,\sqrt{2}\sqrt{\pi}Dirac\left(1,\omega\right)$$

Once more we try it on $f(t) := t^2$, to obtain

> `fouriersin(t^2,t,omega);`

$$-2\,\frac{\sqrt{2}}{\sqrt{\pi}\omega^3}$$

To save typing ω we introduce the `macro`

> `macro(w=omega);`

Let us now use MAPLE to check that $\mathcal{F}_f(\mathcal{F}_f(f)) \equiv f(t)$, that is $\mathcal{F}_f$ is its own inverse.

> `fouriersin(f(t),t,w);`

$$\text{fouriersin}(\text{f}(t),\,t,\,\omega)$$

> `fouriersin(fouriersin(f(t),t,w),w,t);`

$$\text{f}(t)$$

It is easy to check by hand calculation that

$$\mathcal{F}_s\left(\frac{\partial f(x)}{\partial x}; x \to \xi\right) = \xi \mathcal{F}_c(\xi),$$

and

$$\mathcal{F}_c\left(\frac{\partial f(x,y)}{\partial x}; x \to \xi\right) = -f(0,y) + \xi \mathcal{F}_s(\xi,y).$$

Let us check this with MAPLE.

> `fouriersin(diff(f(t),t),t,w);`

$$-\omega\,\text{fouriercos}(\text{f}(t),\,t,\,\omega)$$

Indeed, let us find the transformation rule for the second derivative.

> `fouriersin(diff(f(t),t$2),t,w);`

$$-\frac{\omega\left(\omega\,fouriersin\left(f\left(t\right),t,\omega\right)\sqrt{\pi} - \sqrt{2}f\left(0\right)\right)}{\sqrt{\pi}}$$

Now we check that the Fourier cosine transform is its own inverse.

> `fouriercos(f(t),t,w);`

$$\text{fouriercos}(\text{f}(t),\,t,\,\omega)$$

> `fouriercos(fouriercos(f(t),t,w),w,t);`

$$\text{f}(t)$$

> `fouriercos(diff(f(t),t),t,w);`

$$\frac{\omega\,fouriersin\left(f\left(t\right),t,\omega\right)\sqrt{\pi} - \sqrt{2}f\left(0\right)}{\sqrt{\pi}}$$

We obtain the Fourier cosine transformation law for the derivative of a function.

```
>   fouriercos(diff(f(t),t$2),t,w);
```

$$-\frac{\omega^2 \textit{fouriercos}\left(f\left(t\right),t,\omega\right)\sqrt{\pi}+\sqrt{2}D\left(f\right)(0)}{\sqrt{\pi}}$$

```
>   fourier(f(t),t,w);
```

$$\text{fourier}(f(t),\, t,\, \omega)$$

Let us apply the Fourier transform twice to see what we obtain.

```
>   fourier(fourier(f(t),t,w),w,t);
```

$$2\,\pi\,\text{f}(-t)$$

Hence this proves that indeed the inverse Fourier transform is given by

$$f(x) := \mathcal{F}^{-1}(F(\xi),\xi\to x) := \frac{1}{2\pi}\int\limits_{-\infty}^{\infty} f(\xi)\exp\xi x\,d\xi.$$

However, if we use the command **invfourier** we oftain the original function before Fourier transforming it.

```
>   invfourier(fourier(f(t),t,w),w,t);
```

$$\text{f}(t)$$

The Fourier transform also recognizes the Dirac delta-function.

```
>   fourier(Dirac(t-a),t,w);
```

$$e^{(-I\,a\,\omega)}$$

The Fourier transform will also recognize the Heaviside function.

```
>   fourier(Heaviside(t-a),t,w);
```

$$e^{(-I\,a\,\omega)}\left(\pi\,\text{Dirac}(\omega)-\frac{I}{\omega}\right)$$

```
>   F(w):=fourier(f(t),t,w);
```

$$F := w \mapsto \textit{fourier}\left(f\left(t\right),t,w\right)$$

```
>   diff(F(w),w);
```

$$\frac{\partial}{\partial\omega}\,\text{fourier}(f(t),\, t,\, \omega)$$

```
>   -I*fourier(t*f(t),t,w);
```

$$\frac{\partial}{\partial\omega}\,\text{fourier}(f(t),\, t,\, \omega)$$

```
>   fourier(t^2*f(t),t,w);
```

$$-(\frac{\partial^2}{\partial\omega^2}\,\text{fourier}(f(t),\,t,\,\omega))$$

We continue with the MAPLE package `inttrans`.

> `with(inttrans):`

We define the Fourier transform as $F(s)$ and discover some general properties.

> `F(s):=fourier(f(t),t,s);`

$$F(s) := \text{fourier}(f(t),\,t,\,s)$$

If we take the Fourier transform of $F(s)$ we derive a method of inverting the Fourier transform.

$$F := s \mapsto \textit{fourier}\,(f\,(t)\,,t,s)$$

$$2\,\pi\,f(-t)$$

What is the Fourier transform of $f(t)$ multiplied by the term $\exp i\omega\,t$?

> `fourier(f(t)*exp(I*omega*t),t,s);`

$$\text{fourier}(f(t)\,e^{(I\,\omega\,t)},\,t,\,s)$$

Let us check the Fourier transform of a derivative of a general function $f(t)$.

> `fourier(diff(f(t),t$4),t,s);`

$$s^4\,\text{fourier}(f(t),\,t,\,s)$$

Indeed, we may check the first six derivatives to guess the general formula.

> `for k from 1 to 6 do fourier(diff(f(t),t$k),t,s) od;`

$$I\,s\,\text{fourier}(f(t),\,t,\,s)$$

$$-s^2\,\text{fourier}(f(t),\,t,\,s)$$

$$-I\,s^3\,\text{fourier}(f(t),\,t,\,s)$$

$$s^4\,\text{fourier}(f(t),\,t,\,s)$$

$$I\,s^5\,\text{fourier}(f(t),\,t,\,s)$$

$$-s^6\,\text{fourier}(f(t),\,t,\,s)$$

Hence we surmise that the general expression for the Fourier transform of $\frac{d^k\,f(t)}{dt^k}$ is $(is)^k\mathcal{F}(f)(s)$. It is not difficult to see from the MAPLE expression that the Fourier transform of $f(t) := \frac{1}{t^2+a^2}$

> `assume(a,positive);`

> `fourier(1/(t^2+a^2),t,s);`

$$\frac{\pi\,(e^{-as}\,\textit{Heaviside}\,(s)+e^{as}\,\textit{Heaviside}\,(-s))}{a}$$

$$\frac{\pi\,(e^{-as}\,\textit{Heaviside}\,(s)+e^{as}\,\textit{Heaviside}\,(-s))}{a}$$

has the concise form $\frac{\pi}{a}\exp -a|s|$. We next compute the fourier transform of $\frac{t}{t^2+a^2}$.

```
> fourier(t/(t^2+a^2),t,s);
```

$$i\pi\left(-e^{-as}\,\text{Heaviside}\,(s)+e^{as}\,\text{Heaviside}\,(-s)\right)$$

to be $-i\pi\text{sgn}(s)\exp -a|s|$. We next take the Fourier transforms of some functions which occur in signal analysis. Such a function is the single tooth pulse which we define as

```
> pulse:=proc(t,T)
> Heaviside(t+T)-Heaviside(t-T) end proc;
```

$$pulse := \mathbf{proc}(t,\,T)\,\text{Heaviside}(t+T)-\text{Heaviside}(t-T)\,\mathbf{end}$$

The Fourier transform of this function is

```
> fourier(pulse(t,T),t,s);
```

$$e^{(I\,T\,s)}\left(\pi\,\text{Dirac}(s)-\frac{I}{s}\right)-e^{(-I\,T\,s)}\left(\pi\,\text{Dirac}(s)-\frac{I}{s}\right)$$

```
> tooth:=proc(t,T)
> (1-t/T)*(Heaviside(t+T)-Heaviside(t-T)) end;
```

$$tooth := \mathbf{proc}(t,\,T)\,(1-t/T)\times(\text{Heaviside}(t+T)-\text{Heaviside}(t-T))\;\mathbf{end}$$

We graph the single saw-tooth wave

```
> plot(tooth(t,1),t=-2..2);
```

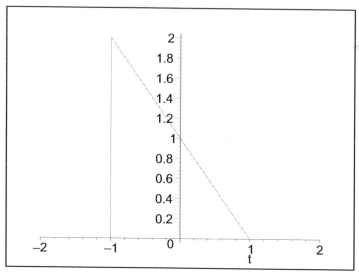

and we compute the Fourier transform

```
> fourier(tooth(t,T),t,s);
```

$$2\,i\left(-\frac{e^{iTs}}{s}+\frac{\sin\,(Ts)}{Ts^2}\right)$$

We continue with our constructing a library of Fourier transforms. The first function is the constant function $f(t) := 1$. We obtain the Dirac delta function

```
>   fourier(1,t,s);
```
$$2\,\pi\,\mathrm{Dirac}(s)$$

Next we seek the general formula for the Fourier transform of t^n.

```
>   for n from 1 to 5 do fourier(t^n,t,s) od;
```
$$2\,I\,\pi\,\mathrm{Dirac}(1,\,s)$$
$$-2\,\pi\,\mathrm{Dirac}(2,\,s)$$
$$-2\,I\,\pi\,\mathrm{Dirac}(3,\,s)$$
$$2\,\pi\,\mathrm{Dirac}(4,\,s)$$
$$2\,I\,\pi\,\mathrm{Dirac}(5,\,s)$$

Hence, we would surmise that $\mathcal{F}(t^n)(s) = 2\pi(i)^n \frac{d^n\,\delta(s)}{ds^n}$.

We next seek the general formula for the Fourier transform of powers of t.

```
>   for n from 1 to 5 do fourier(1/t^n,t,s) od;
```
$$i\pi\,(1 - 2\,\mathit{Heaviside}\,(s))$$
$$\pi\,s\,(\mathrm{Heaviside}(-s) - \mathrm{Heaviside}(s))$$
$$fourier\left(t^{-2}, t, s\right)$$
$$fourier\left(t^{-3}, t, s\right)$$
$$fourier\left(t^{-4}, t, s\right)$$
$$fourier\left(t^{-5}, t, s\right)$$

We surmise the formula is $\mathcal{F}(t^{-n})(s) = i\pi\frac{s^{n-1}}{(n-1)!}\mathrm{sgn}(s)$. Another function for which it is useful to have a Fourier transform for is $f(t) := |t|$.

```
>   fourier(abs(t),t,s);
```
$$-2\,s^{-2}$$

We now try to solve several ordinary differential equations eith the Fourier transform. We consider a differential equation we already did with the Laplace transform and compare the answers.

```
>   ode1:=diff(y(t),t$2)+y(t)=sin(t);
```
$$ode1 := \left(\frac{\partial^2}{\partial t^2}\,y(t)\right) + y(t) = \sin(t)$$

```
>   fourier(ode1,t,omega);
```
$$-(\omega - 1)\,(\omega + 1)\,fourier\,(y\,(t)\,, t, \omega) = i\pi\,(\mathit{Dirac}\,(\omega + 1) - \mathit{Dirac}\,(\omega - 1))$$

```
>   solve(%,fourier(y(t),t,omega));
```
$$\frac{-i\pi\,(\mathit{Dirac}\,(\omega + 1) - \mathit{Dirac}\,(\omega - 1))}{(\omega - 1)\,(\omega + 1)}$$

```
>  soln:=invfourier(-I*Pi*(-Dirac(omega-1)+Dirac(omega+1))/
>  (omega^2-1),omega,t);
```

$$soln := -1/2\, t \cos(t) + 1/4 \sin(t)$$

```
>  convert(soln,trig);
```

$$\frac{1}{8}\, I \left(2\, I \left(\cos(t) + I \sin(t)\right) t - 2\, I \sin(t) + 2\, I \left(\cos(t) - I \sin(t)\right) t\right)$$

```
>  normal(%);
```

$$-\frac{1}{2}\, t \cos(t) + \frac{1}{4} \sin(t)$$

which is a particular solution to the differential equation not the general solution. Why?

Let us try **fourier** on the Bessel equation.

```
>  readlib(fourier):
>  ode3:=t*diff(y(t),t$2)+diff(y(t),t)+t*y(t)=0;
```

$$ode3 := t \left(\frac{\partial^2}{\partial t^2}\, y(t)\right) + \left(\frac{\partial}{\partial t}\, y(t)\right) + t\, y(t) = 0$$

```
>  fourier(ode3,t,omega);
```

$$i\left(-\omega + 1\right)\left(\omega + 1\right)\frac{\partial}{\partial \omega} fourier\left(y\left(t\right), t, \omega\right) - i\omega\, fourier\left(y\left(t\right), t, \omega\right) = 0$$

```
>  subs(fourier(y(t),t,omega)=Y(omega),%);
```

$$i\left(-\omega + 1\right)\left(\omega + 1\right)\frac{d}{d\omega} Y\left(\omega\right) - i\omega\, Y\left(\omega\right) = 0$$

```
>  dsolve(%,Y(omega));
```

$$Y(\omega) = \frac{\_C1}{\sqrt{\omega - 1}\,\sqrt{\omega + 1}}$$

invfourier cannot invert this so we make use of the fact that the Laplace transform and the Fourier transform are related by the change of parameter $\omega = i\,s$.

```
>  subs(omega=I*s,_C1/(omega-1)^(1/2)/(omega+1)^(1/2));
```

$$\frac{\_C1}{\sqrt{I\,s - 1}\,\sqrt{I\,s + 1}}$$

This is the same function we got with the Laplace transform, so we can invert it.

```
>  readlib(laplace):
>  invlaplace(_C1/(s^2+1)^(1/2),s,t);
```

$$\_C1\, \text{BesselJ}(0,\, t)$$

The Fourier transform is most useful for solving partial differential equations. Let us try it on Laplace's equation, where the region we wish to solve for is the half-plane, $y \geq 0$. We first input the Laplacian of the function $u(x, y)$.

```
>  Delta_u:=diff(u(x,y),x$2)+diff(u(x,y),y$2);
```

$$Delta\_u := (\frac{\partial^2}{\partial x^2} u(x, y)) + (\frac{\partial^2}{\partial y^2} u(x, y))$$

Now let us take the Fourier transform of

$$\frac{\partial^2}{\partial x^2} u(x, y)) + (\frac{\partial^2}{\partial y^2} u(x, y)$$

which leads to an ordinary differential equation in the y variable.

```
>   fourier(Delta_u,x,xi);
```

$$-\xi^2 \, \text{fourier}(u(x, y), x, \xi) + (\frac{\partial^2}{\partial y^2} \text{fourier}(u(x, y), x, \xi))$$

To simplify working with these expressions we replace the Fourier transform of $u(x,y)$ bu $U(\xi,y)$.

```
>   de1:=subs(fourier(u(x,y),x,xi)=U(xi,y),%);
```

$$de1 := -\xi^2 \, U(\xi, y) + (\frac{\partial^2}{\partial y^2} U(\xi, y))$$

As $U(\xi,y)$ is a function of **two** variables MAPLE does not want to use dsolve and insists we use pdsolve instead.

```
>   pdsolve(de1,U(xi,y));
```

$$U(\xi, y) = \_F1(\xi) \, e^{(-\xi y)} + \_F2(\xi) \, e^{(\xi y)}$$

If we have Dirichlet data for the Laplace equation , i.e.

$$u(x,0) \; = \; f(x),$$

with the limiting condition

$$u(x, y) \to 0, \sqrt{x^2 + y^2} \to \infty,$$

these conditions imply

$$U(\xi, y) \; = \; F(\xi) \; := \; \mathcal{F}(f(x); x \to \xi),$$

and

$$U(\xi, y) \to 0, \text{ when } y \to \infty.$$

This forces us to chose the Fourier transformed solution to be of the form

$$U(\xi, y) \; = \; F(\xi) \exp -|\xi| y.$$

The solution is then found by computing the inverse Fourier transform; however, without a specific function for $f(x)$ MAPLE just echos back our command.

```
>  invfourier(F(xi)*exp(-abs(xi)*y),xi,x);
```

$$invfourier\left(fourier\left(f\left(t\right),t,\xi\right)e^{-|\xi|y},\xi,x\right)$$

As an example, let us set $F(\xi) := 1$; hence, we compute

```
>  invfourier(exp(-abs(xi)*y),xi,x);
```

$$\text{invfourier}(e^{(-|\xi|\,y)},\xi,x)$$

```
>  assume(y>0);
>  fourier(y/(x^2+y^2),x,xi);
```

$$\pi\left(Heaviside\left(-\xi\right)e^{\xi\,y}+Heaviside\left(\xi\right)e^{-\xi\,y}\right)$$

```
>  simplify(%);
```

$$-\left(Heaviside\left(-\xi\right)e^{\xi\,y}+Heaviside\left(\xi\right)e^{-\xi\,y}\right)$$

```
>  invfourier(%,xi,x);
```

$$-\frac{\tilde{y}}{(\tilde{y}+I\,x)\,(-\tilde{y}+I\,x)}$$

Now let us find an $F(\xi)$ and hence a solution corresponding a reasonable initial function, say $f(x) := 1$.

```
>  F(xi):=fourier(1,x,xi);
```

$$F(\xi) := 2\,\pi\,\mathrm{Dirac}(\xi)$$

Actually it is known that if a Fourier transformed function has the form

$$U(\xi) \ = \ F(\xi)G(\xi),$$

where

$$F(\xi) \ = \ \mathcal{F}(f(x); x \to xi), \ \text{and} \ F(\xi) \ = \ \mathcal{F}(f(x); x \to xi),$$

then u may be written as a **convolution** integral

$$u(x) \ = \ \frac{1}{2\pi}\int\limits_{-\infty}^{\infty} f(\xi)g(x-\xi)\,d\xi.$$

With this is mind we provide a procedure for doing this convolution integral.

```
>  convolution:=proc(f,g,x,y) description
>  "This program computes the convolution integral of two functions,
>  f(x) and g(x)" ;u(t,y):=subs(x=t,f); v(x-t,y):=subs(x=x-t,g);
>  int(u(t,y)*v(x-t,y),t=-infinity..infinity)
>  end proc;
```

convolution := **proc**(*f*, *g*, *x*, *y*)

description

"This program computes the convolution integral of two functions, f(x) and g(x)";

    u(*t*, *y*) := subs(*x* = *t*, *f*);

    v(*x* − *t*, *y*) := subs(*x* = *x* − *t*, *g*);

    int(u(*t*, *y*) * v(*x* − *t*, *y*), *t* = −∞..∞)

end proc

We use this procedure to evaluate the transform.

>   `convolution(1,-y/(y^2+x^2),x,y);`

$$i \ln \left(\frac{i}{y} \right) - i \ln \left(\frac{-i}{y} \right)$$

For the special case we have discussed the solution can then be written as

>   `convolution(1,-y/(y^2+x^2),x,y);`

$$-\pi$$

$$\pi$$

We now consider the boundary value problem associated with Laplace's equation for a strip, $0 \le y \le a$. The boundary data is taken to be

$$u(x,0) = 1, \text{and } x(x,a) = 0.$$

The transformed Laplace equation is as before

>   `deq2:=diff(U(xi,y),y$2)-xi^2*U(xi,y)=0;`

$$deq2 := (\frac{\partial^2}{\partial y^2} U(\xi, y)) - \xi^2 U(\xi, y) = 0$$

First we input the fact that a is positive.

>   `assume(a>0);`

>   `sol:=_F2(xi)*exp(xi*y)+_F1(xi)*exp(-xi*y);`

$$sol := \_F2(\xi) \, e^{(\xi y)} + \_F1(\xi) \, e^{(-\xi y)}$$

We next input the Dirichlet data.

>   `f(x):=1;g(x):=0;`

$$f(x) := 1$$

$$g(x) := 0$$

Next we transform the Dirichlet data.

>   `fourier(f(x),x,xi);`

$$2 \pi \, \text{Dirac}(\xi)$$

```
>    fourier(g(x),x,xi);
```
$$0$$
```
>    subs(y=0,sol)=2*Pi*Dirac(xi);
```
$$\_F2(\xi)\,e^0 + \_F1(\xi)\,e^0 = 2\,\pi\,\mathrm{Dirac}(\xi)$$
```
>    subs(y=a,sol)=0;
```
$$\_F2(\xi)\,e^{(\xi\,a^-)} + \_F1(\xi)\,e^{(-\xi\,a^-)} = 0$$

Next we determine the functions $F_1(\xi)$ and $F_2(\xi)$

```
>    solve({1.*_F2(xi)+1.*_F1(xi) -2*
>    Pi*Dirac(xi),
>    _F2(xi)*exp(xi*a)+_F1(xi)*exp(-1.*xi*a)},{_F1(xi),_F2(xi)});
```

$$\left\{ \_F1(\xi) = -6.283185307\,\frac{Dirac\,(\xi)\ e^{\xi\,a}}{-1.0\,e^{\xi\,a} + e^{-1.0\,\xi\,a}},\ \_F2(\xi) = 6.283185307\,\frac{Dirac\,(\xi)\,e^{-1.0\,\xi\,a}}{-1.0\,e^{\xi\,a} + e^{-1.0\,\xi\,a}} \right\}$$

To obtain the solution to the boundary value problem we need to compute the inverse Fourier transform. We obtain an expected result. Check this result by confirming that it satisfies Laplace's equation and the boundary conditions.

```
>    invfourier(-6.283185307*Dirac(xi)*exp(xi*a)/(-1.*exp(xi*a)
>    +exp(-1.*xi*a))*exp(-xi*y) +
>    6.283185307*Dirac(xi)*exp(-1.*xi*a)/(-1.*exp(xi*a)
>    +exp(-1.*xi*a))*exp(xi*y) ,xi,x);
```

$$0.9999999999\,\frac{a - 1.0\,y}{a}$$

Let us try to use Fourier transforms to solve Cauchy's problem for the heat differential equation. namely

```
>    deq3:=diff(u(x,t),t)-diff(u(x,t),x$2)=0;
```

$$deq3 := \left(\tfrac{\partial}{\partial t}\,\mathrm{u}(x,\,t)\right) - \left(\tfrac{\partial^2}{\partial x^2}\,\mathrm{u}(x,\,t)\right) = 0$$

As is our custom, we Fourier transform the differential equation.

```
>    fourier(deq3,x,omega);
```

$$\omega^2 fourier\,(u\,(x,t)\,,x,\omega) + \frac{\partial}{\partial t} fourier\,(u\,(x,t)\,,x,\omega) = 0$$

```
>    subs(fourier(u(x,t),x,omega)=U(omega,t),%);
```

$$\omega^2\,\mathrm{U}(\omega,\,t) + \left(\tfrac{\partial}{\partial t}\,\mathrm{U}(\omega,\,t)\right) = 0$$

As this is a function depending on two variables ω and t we need to use pdsolve.

```
>    pdsolve(%,U(omega,t));
```

$$\mathrm{U}(\omega,\,t) = \_F1(\omega)\,e^{(-\omega^2\,t)}$$

For the initial (Cauchy) data we choose

$$u(x,0) \;=\; \sin(x);$$

then the Fourier transformed data is

```
>    fourier(sin(x),x,omega);
```

$$i\pi\,(-Dirac\,(\omega - 1) + Dirac\,(\omega + 1))$$

```
>   invfourier((
>   -I*Pi*Dirac(omega-1)+I*Pi*Dirac(omega+1)
>   )*exp(-omega^2*t),omega,x);
```

$$\sin{(x)}\,\mathrm{e}^{-t}$$

We convert the answer to trigonometric form.

```
>   convert(%,trig);
```

$$(\cosh(t) - \sinh(t))\sin(x).$$

Chapter 11

Partial Differential Equations

11.1 Elementary Methods

In this section we solve several elementary partial differential equations of mathematical physics. We start with the potential or Laplace's equation in two space variables. We make use of the MAPLE command pdsolve which frequently finds the general form of a solution. For first-order equations we will see that this may allow us to find a solution to the particular problem we are attempting to solve.

The purpose of this section is to show how we can obtain some rather useful information about solving partial differential equations simply by using the package PDEtools. So first we load this package using the syntax.

```
>  with(PDEtools,build);
```

$$[build]$$

It will turn out that the command pdsolve will frequently give us the general form of the solution of the partial differential equation in question. We first consider first-order wave equations which describe disturbances traveling in one direction with velocity $c(x, t)$.

$$\frac{\partial u}{\partial x} - \frac{1}{c(x,t)} \frac{\partial u}{\partial t} = 0.$$

```
>  pde2:=diff(v(x,t),x)-x*diff(v(x,t),t);
```

$$pde2 := (\tfrac{\partial}{\partial x} v(x,\, t)) - x\,(\tfrac{\partial}{\partial t} v(x,\, t))$$

Applying pdsolve to the above equation yields a general form for the solution.

```
>  pdsolve(pde2,v(x,t));
```

$$v(x,\, t) = \_F1(\frac{x^2}{2} + t)$$

The Cauchy problem is natural for the first-order wave equation type, namely we seek solutions where the initial data is given in the form

$$v(x, 0) = f(x),$$

or more generally in the form

$$v(x(s), y(s)) = \varphi(s),$$

where $(x(s), y(s))$ is the parametric curve for an **initial curve**[1]

In the case of Cauchy condition of the form $v(x, 0) = f(x)$, the initial value problem can be solved by setting

$$F1(x) := f(\sqrt{2x}), \text{ where } x \geq 0.$$

Note that unless $f(x)$ is not an even function of x this solution is not defined for $x < 0$. We consider another first-order wave equation

> `pde3:=diff(w(x,t),x)-x^2*diff(w(x,t),t);`

$$pde3 := (\tfrac{\partial}{\partial x} w(x, t)) - x^2 (\tfrac{\partial}{\partial t} w(x, t))$$

> `pdsolve(pde3,w(x,t));`

$$w(x, t) = \_F1(\frac{x^3}{3} + t)$$

The Cauchy problem for this equation is solved by setting

$$F1(x) := f((3x)^{\frac{1}{3}}), \quad -\infty, x < \infty.$$

We consider next a somewhat more complicated wave equation of the form

> `pde4:=x*diff(w(x,t),x)+t*diff(w(x,t),t)+w(x,t)=0;`

$$pde4 := x (\tfrac{\partial}{\partial x} w(x, t)) + t (\tfrac{\partial}{\partial t} w(x, t)) + w(x, t) = 0$$

> `pdsolve(pde4,w(x,t));`

$$w(x, t) = \frac{1}{x} \_F1 \left(\frac{t}{x} \right)$$

The above equation is not, in general, suitable for the Cauchy problem as $w(x, 0) = \_F1(\frac{0}{x})$ A more natural problem for this equation would be where the solution is prescribed on the line $t = t(x)$ and $w(x, t(x)) = f(x)$. Another wave equation of the above type is

> `pde5:=x*diff(w(x,t),x)+t*diff(w(x,t),t)+x*t*w(x,t)=0;`

$$pde5 := x (\tfrac{\partial}{\partial x} w(x, t)) + t (\tfrac{\partial}{\partial t} w(x, t)) + x t w(x, t) = 0$$

> `pdsolve(pde5,w(x,t));`

$$w(x, t) = \_F1(\frac{t}{x}) e^{(-1/2 x t)}$$

> `pde6:=t*diff(w(x,t),x)+x*diff(w(x,t),t)+w(x,t)=0;`

$$pde6 := t (\tfrac{\partial}{\partial x} w(x, t)) + x (\tfrac{\partial}{\partial t} w(x, t)) + w(x, t) = 0$$

---

[1]An initial curve must not be **characteristic**, that is $\frac{dx}{ds} \neq c(x(s), y(s))$.

```
>   pdsolve(pde6,w(x,t));
```

$$w\left(x,t\right) = -\frac{F1\left(t^2 - x^2\right)}{x+t}$$

The Cauchy problem for the above equation is solved by setting $F1(x) := \sqrt{-x}f(\sqrt{-x})$ which makes sense only if $f(x)$ is an odd function.

We now consider several second-order equations, the first being the heat equation otherwise known as the diffusion equation. In one space dimension it has the form

$$K\frac{\partial^2 u}{\partial x^2} = \frac{\partial u}{\partial t}.$$

We consider the case where the diffusivity $K = 1$.

```
>   pde7:=alpha^2*diff(v(x,t),x$2)-diff(v(x,t),t);
```

$$pde7 := K\left(\frac{\partial^2}{\partial x^2} \, \mathrm{v}(x,\,t)\right) - \left(\frac{\partial}{\partial t}\,\mathrm{v}(x,\,t)\right)$$

```
>   pdsolve(pde7,v(x,t));
```

$$(\mathrm{v}(x,\,t) = \_F1(x)\,\_F2(t)) \,\&\mathrm{where}\,[\{\tfrac{\partial^2}{\partial x^2}\,\_F1(x) = \_c_1\,\_F1(x),\, \tfrac{\partial}{\partial t}\,\_F2(t) = K\,\_c_1\,\_F2(t)\}]$$

We consider next the potential equation, also known as Laplace's equation. As earlier we try `pdsolve` on this second-order equation. We find that we can write the solution as the sum of two *analytic* functions.

```
>   pde8:=diff(v(x,y),x$2)+diff(v(x,y),y$2);
```

$$pde4 := \left(\frac{\partial^2}{\partial x^2}\,\mathrm{v}(x,\,y)\right) + \left(\frac{\partial^2}{\partial y^2}\,\mathrm{v}(x,\,y)\right)$$

```
>   pdsolve(pde8,v(x,y));
```

$$\mathrm{v}(x,\,y) = \_F1(y - I\,x) + \_F2(y + I\,x)$$

This suggests that one might seek to find solutions in the form of power series of the form

$$\_F(y + I\,x) := \sum_{n=0}^{\infty} a_n\,(y + I\,x)^n$$

For example, since $x + Iy = re^\theta$, a real solution can be written as

$$\sum_{n=0}^{\infty} (a_n\cos(n\theta) + b_n\sin(n\theta))\,r^n$$

Project: This suggests that the Laplace equation for a circle of radius 1 with boundary data $u(L,\theta) = f(\theta)$ might be solved by computing the coefficients a_n, b_n from the formulae

$$a_n\int_0^1 \cos^2(n\theta)\,d\theta = \int\int_0^1 f(\theta)\cos(n\theta)\,d\theta,$$

$$b_n\int_0^1 \sin^2(n\theta)\,d\theta = \int\int_0^1 f(\theta)\cos(n\theta)\,d\theta.$$

Write a procedure which does this.

Our next equation is the second-order wave equation. It differs from the first-order version in that it has two traveling disturbances, each traveling in an opposite direction from the other. The second-order wave equation has the form

$$\frac{\partial^2 u(x,t)}{\partial x^2} - \frac{1}{c^2(x,t)}\frac{\partial^2 u(x,t)}{\partial t^2},$$

where $c(x,t)$ is the wave speed. Let consider the simple case where $c = 1$. Then the equation takes the form

```
>    pde9:=diff(v(x,t),x$2)-diff(v(x,t),t$2);
```

$$pde9 := \left(\frac{\partial^2}{\partial x^2}\, v(x,\,t)\right) - \left(\frac{\partial^2}{\partial t^2}\, v(x,\,t)\right)$$

pdsolve tells us that the solution can be written in terms of two functions representing traveling waves to the left and to the right.

```
>    pdsolve(pde9,v(x,t));
```

$$v(x,\,t) = \_F1(t+x) + \_F2(t-x)$$

The Cauchy problem in this case is to find a solution which satisfies the following initial data

$$u(x,0) = f(x),\ ,\ \text{and}\ \frac{\partial u(x,t)}{\partial t}(x,0) = g(x),$$

where f and g are arbitrary prescribed functions. Clearly $\_F1$ and $\_F2$ may be found from f and g.

Project: Write a program which solves the Cauchy problem for the second-order wave equation.

11.2   The First-Order Partial Differential Equation

In this session we see how to use MAPLE to solve first-order partial differential equations of quasilinear type, namely

$$a(x, y, u)\frac{\partial u}{\partial x} + b(x, y, u)\frac{\partial u}{\partial y} = c(x, y, u).$$

The solution defines a surface $u = u(x, y)$, whose normal is the vector $[u_x, u_y, -1]$. As this normal is perpendicular to the vector $[a, b, c]$, the vector $[a, b, c]$ is tangent to the surface. Hence the characteristics of the differential equation are curves lying on the surface, which are defined by the ordinary differential equations.

$$\frac{dx}{a(x, y, u)} = \frac{dy}{b(x, y, u)} = \frac{du}{c(x, y, u)}.$$

A natural problem to consider for the first-order partial differential equation is the Cauchy problem where the data is prescribed on a non-characteristic curve

(CD) $$x = x(t), \ y = y(t), \ u = u(t).$$

We then try to solve the Cauchy problem by solving the characteristic equations for each point which passes through the non-characteristic curve. If we use the arc-length s as the parameter to measure along the characteristic curves, we construct a surface described in terms of the parameters s and t

$$x = x(s,t), \ y = y(s,t), \ u = u(s,t).$$

To solve the characteristic equations we replace the system of ordinary equations for the characteristics by the system

$$\frac{dx}{ds} = a(x,y,u), \quad \frac{dy}{ds} = b(x,y,u), \quad \frac{du}{ds} = c(x,y,u).$$

We begin our exploration of this method with MAPLE by starting with the quasilinear partial differential equation

```
>   pde1:=diff(u(x,y),x)+diff(u(x,y),y)-u(x,y)^2;
```

$$pde1 := \left(\tfrac{\partial}{\partial x}\, u(x,\,y)\right) + \left(\tfrac{\partial}{\partial y}\, u(x,\,y)\right) - u(x,\,y)^2$$

We try `pdsolve` which provides us with a general solution.

```
>   pdsolve(pde1,u(x,y));
```

$$u(x,\,y) = -\frac{1}{x - \_F1(y+x)}$$

However, let us try the approach sketched above. First we need to load PDEtools. We leave the semi-colon ending on to see what is contained in PDEtools.

```
>   with(PDEtools);
```

$[PDEplot,\ build,\ charstrip,\ dchange,\ difforder,\ mapde,\ splitstrip,\ splitsys]$

In order to calculate the surface we use `charstrip` which produces the system corresponding to (CD) for this equation.

```
>   sys1 := charstrip(pde1,u(x,y));
```

$$\left\{ u(\_s) = (-\_s + \_C1)^{-1},\, x(\_s) = \_s + \_C3,\, y(\_s) = \_s + \_C2 \right\}$$

As this is a system of ordinary equations we use `dsolve` to solve it.

```
>   dsolve(sys1, {u(_s),x(_s),y(_s)}, explicit);
```

$$\left\{ u(\_s) = \frac{1}{-\_s + \_C1},\, x(\_s) = \_s + \_C3,\, y(\_s) = \_s + \_C2 \right\}$$

In order to make this approach useful we need to specify an initial curve in the form (CD). The initial curve corresponds to the arc-length parameter $s = 0$. Hence, let us substitute $s = 0$ into the solution. We Shall then choose the arbitrary "constants" _C1, _C2, _C3 so that as functions of t these "constants" correspond to a point, indexed by t, on the initial curve.

```
>  subs(_s=0,{u(_s) = 1/(-_s+_C1), x(_s) = _s+_C2, y(_s) = _s+_C3});
```

$$\{u(0) = \frac{1}{\_C1}, x(0) = \_C2, y(0) = \_C3\}$$

As an initial curve we choose

```
>  subs({u(0)=_t,x(0)=_t,y(0)=-_t},%);
```

$$\{\_t = \_C2, \_t = \_C1^{-1}, -\_t = \_C3\}$$

```
>  solve({_t = _C2, _t = 1/_C1,-_t = _C3},{_C1,_C2,_C3});
```

$$\{\_C3 = -\_t, \_C2 = \_t, \_C1 = \frac{1}{\_t}\}$$

The we find that the coefficients may be replaced as function of the parameter t

```
>  subs({_C3 = -_t, _C2 = _t, _C1 =
>  1/_t},{u(_s,_t) = 1/(-_s+_C1),
>  x(_s,_t) = _s+_C2, y(_s,_t) = _s+_C3});
```

$$\left\{ u(\_s, \_t) = \frac{1}{-\_s + \frac{1}{\_t}}, x(\_s, \_t) = \_s + \_t, y(\_s, \_t) = \_s - \_t \right\}$$

We now want to write the solution u as a function of x and y. This is done by solving the system

$$x = x(s, t), \quad y = y(s, t)$$

for $s = s(x, y)$, $t = t(x, y)$ and using this to replace s and t in $u(s, t)$. This can always be done, for sufficiently small values of s, providing

$$\frac{\partial x}{\partial s} \frac{\partial y}{\partial t} - \frac{\partial x}{\partial t} \frac{\partial y}{\partial s} \neq 0$$

along the initial curve.

```
>  solve( {x = _s+_t, y = _s-_t},{_s,_t});
```

$$\{\_s = x/2 + y/2, \_t = x/2 - y/2\}$$

```
>  subs({_t = 1/2*x-1/2*y, _s = 1/2*y+1/2*x},u =1/(-_s+1/_t));
```

$$u = \left(-y/2 - x/2 + (x/2 - y/2)^{-1}\right)^{-1}$$

It can be seen that the solution blows up along the hyperbolic curve $x^2 - y^2 = 4$.

```
>  simplify(%);
```

$$u = \frac{-2x + 2y}{x^2 - y^2 - 4}$$

We plot the solution now to exhibit its singular behavior.

```
>  plot3d( -2*(x-y)/(x^2-y^2-4),
```

```
>   x=-10..10,y=-10..10,axes=boxed,numpoints=2000);
>   IC:={u(0) = 1/_C1, x(0) = _C2, y(0) = _C3};
```

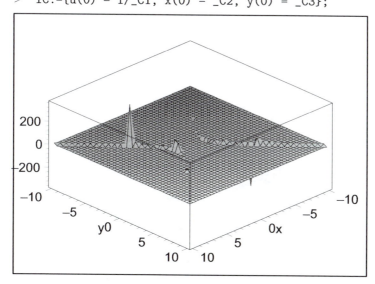

$$IC := \{u(0) = \frac{1}{\_C1}, \, x(0) = \_C2, \, y(0) = \_C3\}$$

We consider the same differential equation but with different initial data

```
>   subs({u(0)=_t^2,x(0)=_t,y(0)=-_t},IC);
```

$$\{\_t = \_C2, \, \_t^2 = \_C1^{-1}, -\_t = \_C3\}$$

```
>   solve({-_t = _C3, _t = _C2, _t^2 = 1/_C1},{_C1,_C2,_C3});
```

$$\{\_C1 = \frac{1}{\_t^2}, \, \_C2 = \_t, \, \_C3 = -\_t\}$$

```
>   subs({_C3 = -_t, _C2 = _t, _C1 =
>   1/_t^2},{u = 1/(-_s+_C1), x =
>   _s+_C2, y = _s+_C3});
```

$$\left\{ u = \frac{1}{-\_s + \dfrac{1}{\_t^2}}, \, x = \_s + \_t, \, y = \_s - \_t \right\}$$

```
>   solve({y = _s-_t,  x = _s+_t},{_s,_t});
```

$$\{\_s = x/2 + y/2, \, \_t = x/2 - y/2\}$$

We obtain a different solution.

```
>   subs({_t = 1/2*x-1/2*y, _s =
>   1/2*y+1/2*x},u  =
>   1/(-_s+1/(_t^2)));
```

$$u = \left(-x/2 - y/2 + (x/2 - y/2)^{-2} \right)^{-1}$$

```
>  simplify(%);
```

$$u = -\frac{2(x-y)^2}{x^3 - x^2\,y - x\,y^2 + y^3 - 8}$$

```
>  plot3d(-2(x-y)^2/(x^3-x^2*y-x*y^2+y^3-8),
>  x=-4..4,y=-4..4,numpoints=2000,axes=boxed);
```

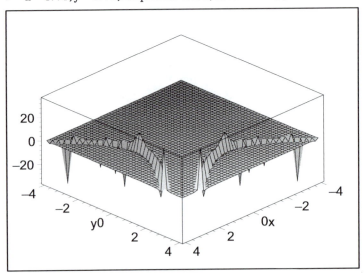

As another example we consider the partial differential equation

```
>  pde3:=x*diff(u(x,y),x)+y*diff(u(x,y),y)-m*u(x,y);
```

$$pde3 := x\left(\tfrac{\partial}{\partial x}\,\mathrm{u}(x,\,y)\right) + y\left(\tfrac{\partial}{\partial y}\,\mathrm{u}(x,\,y)\right) - m\,\mathrm{u}(x,\,y)$$

We use `charstrip` to obtain the characteristic equations

```
>  sys3 := charstrip(pde3,u(x,y));
```

$$sys3 := \left\{\frac{\mathrm{d}}{\mathrm{d}\_s}u(\_s) = mu(\_s),\,\frac{\mathrm{d}}{\mathrm{d}\_s}x(\_s) = x(\_s),\,\frac{\mathrm{d}}{\mathrm{d}\_s}y(\_s) = y(\_s)\right\}$$

and we solve this system of ordinary differential equations with `dsolve`

```
>  dsolve(sys3, {u(_s),x(_s),y(_s)}, explicit);
```

$$\{u(\_s) = \_C3\,\mathrm{e}^{m\_s},\, x(\_s) = \_C2\,\mathrm{e}^{-s},\, y(\_s) = \_C1\mathrm{e}^{-s}\}$$

We next try to fit the characteristic curves to the initial curve.

```
>  subs(_s=0,{x(_s) = exp(_s)*_C1, y(_s) =
>  exp(_s)*_C2, u(_s) =
>  exp(m*_s)*_C3});
```

$$\{u(0) = \_C3\,\mathrm{e}^0,\, x(0) = \_C1\,\mathrm{e}^0,\, y(0) = \_C2\,\mathrm{e}^0\}$$

```
>  subs({u(0)=_t,x(0)=_t,y(0)=a*_t^2},%);
```

$$\{\_t = \_C1,\, a\,\_t^2 = \_C2,\, \_t = \_C3\}$$

```
> solve({_t = _C3, a*_t^2 = _C2, _t = _C1},{_C1,_C2,_C3});
```

$$\{ \_C1 = \_t, \ \_C2 = a\_t^2, \ \_C3 = \_t\}$$

We replace the constants _C1, _C2,_C3 by functions of t, which yields the surface parametrically in terms of s and t

```
>  subs({_C1 = _t, _C2 = a*_t^2, _C3 =
>  _t},{x(_s) = exp(_s)*_C1,
>  y(_s) = exp(_s)*_C2, u(_s) = exp(m*_s)*_C3});
```

$$\{u(\_s) = \_t\,e^{m-s}, x(\_s) = \_t\,e^{-s}, y(\_s) = a\_t^2 e^{-s}\}$$

We now solve for s and t in terms of x and y.

```
> solve({x = exp(_s)*_t, y = exp(_s)*a*_t^2},{_s,_t});
```

$$\left\{ \_s = \ln\left(\frac{ax^2}{y}\right), \ \_t = \frac{y}{xa}\right\}$$

```
> simplify(subs({_t = y/(x*a), _s =
> ln(a*x^2/y),m=5,a=1},u =
> exp(m*_s)*_t));
```

$$u = x^5$$

11.3   The Heat Equation

In this session we consider the initial-boundary-value problem associated with the one-dimentional, heat equation

$$K\frac{\partial^2 u}{\partial x^2} = \frac{\partial u}{\partial t}, \quad 0 < x < L, \ t > 0,$$

where K is the diffusivity and is measures in the units $[\text{cm}]^2[\text{sec}]^{-1}$. The method is based on the separation of variables technique, which allows the solution which satisfies the initial condition $u(x,0) = f(x)$, and the boundary conditions $u(0,t) = 0, u(L,t) = 0$ to be written in the form

$$u(x,t) = \sum_{k=1}^{\inf} c_k \exp -k^2\pi^2 Kt/L^2 \sin\left(\frac{k\pi x}{L}\right)$$

As an approximate solution we consider the truncated version of the above series

$$u(x,t) = \sum_{k=1}^{N} c_k \exp\left(\frac{-k^2\pi^2 Kt}{L^2}\right) \sin\left(\frac{k\pi x}{L}\right)$$

We write this representation as a procedure, where f is the initial data, N the number of terms in the truncated solution, and T the length of the time interval $[0, T]$ over which we want to find the solution. The procedure computes the Fourier coefficients of the initial data f using the formula

$$c_k = \frac{2}{L} \int_0^L f(x) \sin\left(\frac{k\pi x}{L}\right) \, dx$$

```
>  heat_equation_1:=proc(alpha,f,N,x,L,t,T)
>  description "This program uses the separation of variables method
>  to plot the solution of the heat equation in a rod, where the ends
>  of the rod are kept at 0 degrees.  The solution is truncated after
>  N terms." ;local k: for k from 1 to N do
>  c(k):=2/L*int(f*sin(k*Pi*x/L), x=0..L) od; plot3d(
>  sum(c(j)*exp(-j^2*Pi^2*alpha^2*t/L^2)*sin(j*Pi*x/L),j=1..N)
>  ,x=0..L,t=0..T,axes=BOXED) end;
```

$heat\_equation\_1 := \textbf{proc}(\alpha, f, N, x, L, t, T)$

local k;

description "This program uses the separation of variables method to plot the solution of the heat equation in a rod, where the ends of the rod are kept at 0 degrees. The solution is truncated after N terms.";

  for k **to** N **do** $c(k) := 2 * \text{int}(f * \sin(k * \pi * x/L), x = 0..L)/L$ **end do**;

  $\text{plot3d}(\text{sum}(c(j) * \exp(-j^2 * \pi^2 * \alpha^2 * t/L^2) * \sin(j * \pi * x/L), j = 1..N),$
  $x = 0..L, t = 0..T, axes = BOXED)$

end proc

```
>  with(plots):
```

We consider several examples. Suppose we consider the problem where the initial distribution of temperature constant, say $u(x, 0) = 100\%$ and we take the diffusivity $K = 0.25$. We consider several truncation indices, namely $N = 10, 20, 40, 100$. The graphs which are produced by our program are shown below.

```
>  heat_equation_1(.25,100,10,x,50,t,2000);
```

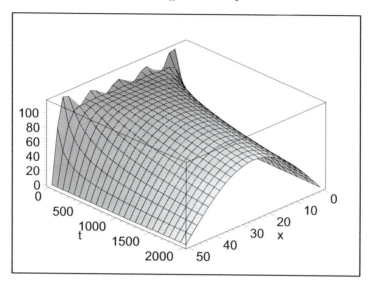

```
>   heat_equation_1(.25,100,20,x,50,t,2000);
```

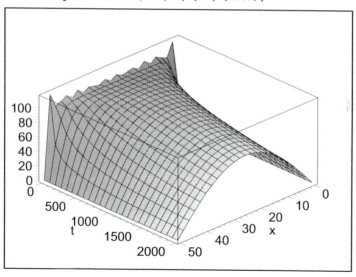

```
>   heat_equation_1(.25,100,40,x,50,t,2000);
```

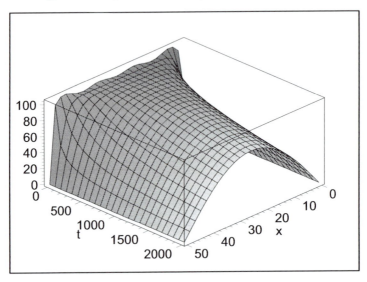

```
>  heat_equation_1(.25,100,100,x,50,t,2000);
```

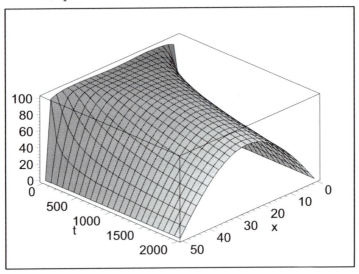

Notice that the heat distribution tends to smooth out to zero because heat flows out of the rod on each end. In future calculations we will take N as 100 as this seems sufficiently large to give good comparison with numerical computations.

We next consider the problem of the temperature distribution of a rod of length L, with insulated ends. The solution is given by

$$u(x,t) = \frac{b_0}{2} + \sum_{n=1}^{\infty} b_n exp\left(-\frac{n^2\pi^2 K t}{L^2}\right) \cos\left(\frac{n\pi x}{L}\right),$$

where the Fourier coefficients are

$$b_n = \frac{2}{L} \int_0^L u(x,0) \cos\left(\frac{n\pi x}{L}\right) dx.$$

The following program is based on this formula.

```
>   heat_equation_2:=proc(alpha,f,N,x,L,t,T)
>   description "This program uses the separation of variables method
>   to plot the solution of the heat equation in a rod, where the ends
>   of the rod are insulated.  The solution is truncated after N
>   terms." ;local k: for k from 0 to N do
>   b(k):=2/L*int(f*cos(k*Pi*x/L), x=0..L) od; plot3d(
>   b(0)/2+sum(b(j)*exp(-j^2*Pi^2*alpha^2*t/L^2)*cos(j*Pi*x/L),j=1..N)
>   ,x=0..L,t=0..T,axes=BOXED) end;
```

$heat\_equation\_2 := \mathbf{proc}(\alpha,\, f,\, N,\, x,\, L,\, t,\, T)$
local k;
description "This program uses the separation of variables method to plot the solution of the heat equation in a rod, where the ends of the rod are insulated. The solution is truncated after N terms.";
  for k **from** 0 **to** N **do** $b(k) := 2 * int(f * \cos(k * \pi * x/L),\, x = 0..L)/L$ **end do** ;
  $plot3d(1/2 * b(0) + sum(b(j) * \exp(-j^2 * \pi^2 * \alpha^2 * t/L^2) * \cos(j * \pi * x/L),$
   $j = 1..N),\, x = 0..L,\, t = 0..T,\, axes = BOXED)$
end proc

We consider an initial heat distribution with parabolic form

$$u(x,0) = u(x,0) = \frac{100x(50 - x)}{625}.$$

Notice from the graph that since no heat escapes because of the insulated ends that the heat distribution smoothes out to an average value.

```
>   heat_equation_2(.25,100*x*(50-x)/625,40,x,50,t,2000);
```

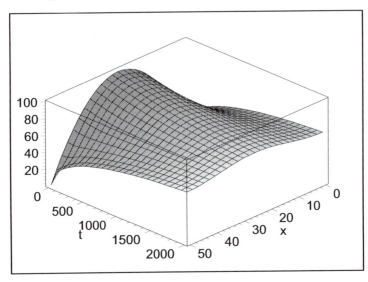

Next we consider a triangular distribution given as

$$f(x) := \begin{cases} 4x & \text{when } 0 \le x < 25 \\ 100 - 4(x - 25) & \text{when } 25 \le x \le 50 \end{cases}$$

To express this in MAPLE we use the Heaviside function which we have defined in Chapter 2.

```
> f:=4*x-8*(x-25)*Heaviside(x-25);
```

$$f := 4\,x - (8x - 100)\,\text{Heaviside}(x - 25)$$

```
> plot(f,x=0..50);
```

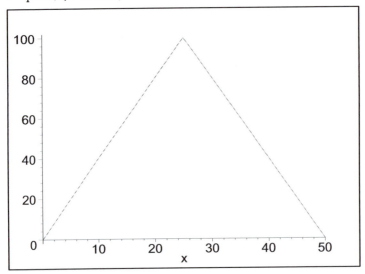

Notice that the initial curve of the surface we have plotted has a sharp point at $x = 25$ but that this smoothes out immediately. This smoothing process is a property of the heat equation. As before the distribution appears to tend to a uniform average temperature.

```
>   heat_equation_2(.25,f,40,x,50,t,2000);
```

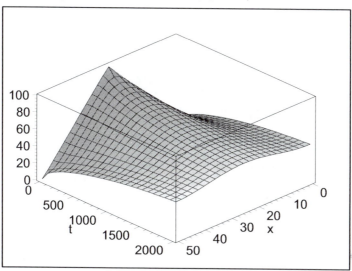

11.4   The Vibrating String

The vibrating string can be modeled as a one dimensional, second-order, wave equation. In this section we consider the boundary-initial-value problem for the string equation. We seek a solution of the boundary-initial value problem for the wave equation

$$\frac{\partial^2 u(x,t)}{\partial x^2} - \frac{1}{a^2}\frac{\partial^2 u(x,t)}{\partial t^2} = 0$$
$$u(0,t) = 0 \qquad u(L,t) = 0$$
$$u(x,0) = f(x) \qquad \frac{\partial u}{\partial t}(x,0) = g(x).$$

First we load DEtools and linalg As we found in Section 1 of this Chapter the separated solutions

$$u(x,t) = X(x)T(t)$$

satisfy the equations.

$$\frac{d^2\,X(x)}{dx^2} = \lambda^2\,X(x) = 0$$
$$\frac{d^2\,T(t)}{dt^2} = \lambda\,a\,T(y) = 0.$$

This suggests that we seek a solution as an infinite series of terms

$$u(x,t) = \sum X_n(x)T_n(t),$$

where the functions $X_n(x)$, $T_n(t)$ must be of the form

$$X_n(x) = \alpha_n \sin(x\lambda_n) + \beta_n \cos(x\lambda_n)$$
$$T_n(t) = \delta_n \sin(a\,t\lambda_n) + \gamma_n \cos(a\,t\lambda_n)$$

The numbers λ_n are the eigen values and are chosen so that $X_n(x)$ satisfy the prescribed boundary conditions. It turns out that for the present case the eigenvalues are given by $\lambda_n = \frac{n\pi}{L}, n = 1, 2, \ldots.$

$$u(x,t) = \sum_{n=1}^{\infty}\left(A_n \cos\left(\frac{n\pi\,a\,t}{L}\right) + B_n \sin\left(\frac{n\pi\,a\,t}{L}\right)\right)\sin\left(\frac{n\pi\,x}{L}\right).$$

Based on this formula we can write a program that gives an approximate solution to the string equation by truncating the series solution after N terms.

```
>   wave_equation_1:=proc(a,f,g,N,x,L,t,T)
>   description "This program uses the separation of variables method
>   to plot the displacement of a string during oscillatory motion.
>   The displacement of the string at its end points is zero.  The
>   solution is truncated after N terms." ;local k, n: for k from 1 to
>   N do A(k):=2/L*int(f*sin(k*Pi*x/L), x=0..L) od; for n from 1 to N
>   do B(n):=2/(n*Pi*a)*int(g*sin(n*Pi*x/L), x=0..L) od;  plot3d(
>   sum((A(j)*cos(j*Pi*a*t/L)+B(j)*sin(j*Pi*a*t/L))*sin(j*Pi*x/L),
>   j=1..N),x=0..L,t=0..T,axes=BOXED) end;
```

wave_equation_1 := **proc**(a, f, g, N, x, L, t, T)
local k, n;
description "This program uses the separation of variables method to plot the displacement of a string during oscilatory motion. The displacement of the string at its end points is zero. The solution is truncated after N terms";
  for k **to** N **do** A(k) $:= 2 * \mathrm{int}(f * \sin(k * \pi * x/L), x = 0..L)/L$ **end do**;
  for n **to** N **do** B(n) $:= 2 * \mathrm{int}(g * \sin(n * \pi * x/L), x = 0..L)/(n * \pi * a)$ **end do** ;
  plot3d(
    sum((A(j) $* \cos(j * \pi * a * t/L)$ + B(j) $* \sin(j * \pi * a * t/L)) * \sin(j * \pi * x/L)$,
    $j = 1..N$), $x = 0..L$, $t = 0..T$, $axes = BOXED$)
end proc

We now test our program by considering the plucked string which we model by choosing the initial displacement to be given by

$$u(x,0) = \begin{array}{ll} x & \text{when } 0 \le x \le \frac{1}{2} \\ 1 - x & \text{when } \frac{1}{2} < x \le 1 \end{array}.$$

The plot below shows that the triangular wave repeats itself in the space-time plot.

```
>   h:=x-2*(x-1/2)*Heaviside(x-1/2);
```

$$h := x - (2x - 1)\,\text{Heaviside}(x - \frac{1}{2})$$

```
>   plot(h,x=0..1);
```

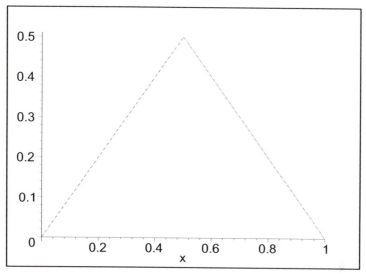

```
>   wave_equation_1(1,h,0,40,x,1,t,200)
>   ;
```

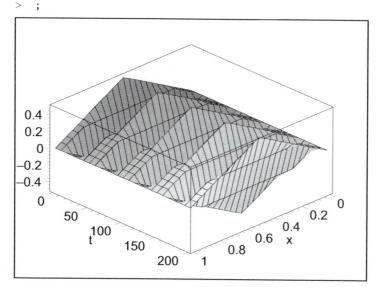

The previous program shows the displacement in space-time. Sometimes it is more useful to see the displacement at certain specified times. The program `snap_shot_string` plots the displacement at prescribed time intervals t_0 for N intervals.

```
>   snap_shot_string:=proc(a,f,g,N,x,L,t_0)
>   description "This program uses the separation of variables method
>   to plot the displacement of a string at specific times during
>   oscilatory motion. The displacement of the string at its end
>   points is zero.  The solution is truncated after N terms." ;local
>   k, n, m: for k from 1 to N do A(k):=2/L*int(f*sin(k*Pi*x/L),
>   x=0..L) od; for n from 1 to N do
>   B(n):=2/(n*Pi*a)*int(g*sin(n*Pi*x/L), x=0..L) od; for m from 1 to
>   9 do G(m):=plot(
>   sum((A(j)*cos(j*Pi*a*m*t_0/L)+B(j)*sin(j*Pi*a*m*t_0/L))*
>   sin(j*Pi*x/L), j=1..N) ,x=0..L) od:
>   display({G(1),G(2),G(3),G(4),G(5),G(6),G(7),G(8),G(9)})
>   end;
```

h

siao1665

```
>   snap_shot_string(1,h,0,10,x,1,Pi/20);
```

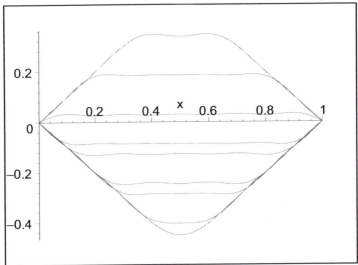

11.4.1   Separation of Variables with MAPLE.

It is actually possible to do the separation of variables with MAPLE. This can be useful when we have a more complicated partial differential equation and the separation process is difficult to perform without the possibility of an algebraic error creeping in. We illustrate the method with the string equation.

```
>   pde:=diff(u(x,t),t$2) - a^2 *
>   diff(u(x,t),x$2)=0;
```

$$pde := (\frac{\partial^2}{\partial t^2} \, \mathrm{u}(x,\,t)) - a^2 \, (\frac{\partial^2}{\partial x^2} \, \mathrm{u}(x,\,t)) = 0$$

As usual, we seek solutions of the form $u(x,t) = X(x)T(t)$
```
>  subs(u(x,t)=X(x)*T(t), pde);
```

$$(\frac{\partial^2}{\partial t^2} \, \mathrm{X}(x)\,\mathrm{T}(t)) - a^2 \, (\frac{\partial^2}{\partial x^2} \, \mathrm{X}(x)\,\mathrm{T}(t)) = 0$$

```
>  normal(%);
```

$$\mathrm{X}(x)\,(\frac{\partial^2}{\partial t^2} \, \mathrm{T}(t)) - a^2 \, (\frac{\partial^2}{\partial x^2} \, \mathrm{X}(x))\,\mathrm{T}(t) = 0$$

We now use MAPLE to extract the separated solutions
```
>   %/(X(x)*T(t));
```

$$\frac{\mathrm{X}(x)\,(\frac{\partial^2}{\partial t^2} \, \mathrm{T}(t)) - a^2 \, (\frac{\partial^2}{\partial x^2} \, \mathrm{X}(x))\,\mathrm{T}(t)}{\mathrm{X}(x)\,\mathrm{T}(t)} = 0$$

```
>  expand(%);
```

$$\frac{\frac{\partial^2}{\partial t^2} \, \mathrm{T}(t)}{\mathrm{T}(t)} - \frac{a^2 \, (\frac{\partial^2}{\partial x^2} \, \mathrm{X}(x))}{\mathrm{X}(x)} = 0$$

```
>  pd:=op(1,%);
```

$$pd := \frac{\frac{\partial^2}{\partial t^2} \, \mathrm{T}(t)}{\mathrm{T}(t)} - \frac{a^2 \, (\frac{\partial^2}{\partial x^2} \, \mathrm{X}(x))}{\mathrm{X}(x)}$$

```
>  -X(x)*(op(2,pd)/a^2-lambda^2)=0;
```

$$-\mathrm{X}(x)\left(-\frac{\frac{\partial^2}{\partial x^2} \, \mathrm{X}(x)}{\mathrm{X}(x)} - \lambda^2\right) = 0$$

Thus we obtain two separated equations
```
>  odeX:=normal(%);
```

$$odeX := \lambda^2 X\,(x) + \frac{d^2}{dx^2} X\,(x) = 0$$

```
>  odeT:=normal(T(t)*(op(1,pd) +
>  a^2*lambda^2)=0);
```

$$odeT := (a^2\,\lambda^2\,\mathrm{T}(t) + \frac{\partial^2}{\partial t^2} \, \mathrm{T}(t)) = 0$$

We next find the general solutions of the separated equations
```
>  solnX:=dsolve(odeX,X(x));
```

$$solnX := X\,(x) = \_C1 \, \sin\left(\frac{n\pi\,x}{l}\right) + \_C2 \, \cos\left(\frac{n\pi\,x}{l}\right)$$

We now seek those separated $X(x)$ solutions which satisfy the boundary conditions

$$X(0) = X(1) = 0.$$

The boundary condition at $x = 0$ forces _$C1$ to be zero.

```
>   subs(x=0,rhs(solnX))=0;
```

$$\_C1 \sin(0) + \_C2 \cos(0) = 0$$

```
>   simplify(%);
```

$$\_C2 = 0$$

The boundary condition at $x = L$ forces the constant λ to be chosen as $\frac{n\pi}{L}$.

```
>   simplify(subs({x=1,_C2=0},lhs(solnX))=0);
```

$$\_C2 \sin(\lambda l) = 0$$

```
>   lambda:=n*Pi/1;
```

$$\lambda := \frac{n\pi}{l}$$

The solutions to the x equation are then of the form

```
>   Xn(x):=coeff(rhs(solnX),_C1);
```

$$\text{Xn}(x) := \sin(\frac{n\pi x}{l})$$

We turn our attention to the equation for $T(t)$.

```
>   solnT:=dsolve(odeT,T(t));
```

$$solnT := \text{T}(t) = \_C1 \sin(\frac{\pi n a t}{l}) + \_C2 \cos(\frac{\pi n a t}{l})$$

```
>   T1n(t):=coeff(rhs(solnT),_C2);
```

$$\text{T1n}(t) := \cos(\frac{\pi n a t}{l})$$

```
>   T2n(t):=coeff(rhs(solnT),_C1);
```

$$\text{T2n}(t) := \sin(\frac{\pi n a t}{l})$$

The separated solutions are then of the form

```
>   u[n](x,t):=T1n(t)*Xn(x);
```

$$u_n(x, t) := \cos(\frac{\pi n a t}{l}) \sin(\frac{n\pi x}{l})$$

```
>   v[n](x,t):=T2n(t)*Xn(x);
```

$$v_n(x,\, t) := \sin(\frac{\pi\, n\, a\, t}{l})\sin(\frac{n\, \pi\, x}{l})$$

```
>   Un(x,t):=factor(a[n]*u[n](x,t) +
>   b[n]*v[n](x,t));
```

$$\mathrm{Un}(x,\, t) := \sin(\frac{n\, \pi\, x}{l})\,(a_n \cos(\frac{\pi\, n\, a\, t}{l}) + \sin(\frac{\pi\, n\, a\, t}{l})\, b_n)$$

```
>   UnP(x,t):=diff(Un(x,t),t);
```

$$UnP := -\frac{a_n\, \pi\, na}{l}\sin\left(\frac{n\pi\, at}{l}\right)\sin\left(\frac{n\pi\, x}{l}\right) + \frac{b_n\, \pi\, na}{l}\cos\left(\frac{n\pi\, at}{l}\right)\sin\left(\frac{n\pi\, x}{l}\right)$$

```
>   simplify(subs(t=0,Un(x,t)));
```

$$a_n \sin\left(\frac{n\pi\, x}{l}\right)$$

```
>   simplify(subs(t=0,UnP(x,t)));
```

$$\frac{b_n\, \pi\, na}{l}\sin\left(\frac{n\pi\, x}{l}\right)$$

Therefore the general form of a solution satisfying the boundary conditions $u(0,t) = u(L,t) = 0$ is given by

```
>   UN(x,t):=sum(Un(x,t),n=1..infinity);
```

$$UN := \sum_{n=1}^{\infty} a_n \cos\left(\frac{n\pi\, at}{l}\right)\sin\left(\frac{n\pi\, x}{l}\right) + b_n \sin\left(\frac{n\pi\, at}{l}\right)\sin\left(\frac{n\pi\, x}{l}\right)$$

11.5   The Laplace Equation

In this session we want to use MAPLE to solve Laplace's equation. First we load the package PDEtools.

```
>   with(PDEtools);
```

$[PDEplot,\ build,\ charstrip,\ dchange,\ difforder,\ mapde,\ splitstrip,\ splitsys]$

Next we enter the two-dimensional Laplace equation

```
>   pde1:=diff(u(x,y),x$2)+diff(u(x,y),y$2)=0;
```

$$pde1 := (\frac{\partial^2}{\partial x^2}\, u(x,\, y)) + (\frac{\partial^2}{\partial y^2}\, u(x,\, y)) = 0$$

When `pdsolve` is applied to this equation we get the family of separated solutions $F_1(x)F_2(y)$ where the F_k satisfy the separated equations.

```
>   pdsolve(pde1,u(x,y));
```

$(\mathrm{u}(x,\, y) = \_F1(x)\, \_F2(y))\, \&where\, [\{\frac{\partial^2}{\partial x^2}\, \_F1(x) = \_c_1\, \_F1(x),\ \frac{\partial^2}{\partial y^2}\, \_F2(y) = -\_c_1\, \_F2(y)\}]$

In order to solve the resulting ordinary differential equations we load DEtools and then use `dsolve` to get their general solutions.

```
>   with(DEtools):
```

> `dsolve(diff(_F1(x),x$2)) = _c[1]*_F1(x),_F1(x));`

$$\_F1(x) = \_C1 \sinh(\sqrt{\_c_1}\, x) + \_C2 \cosh(\sqrt{\_c_1}\, x)$$

> `dsolve(diff(_F2(y),y$2)) = -_c[1]*_F2(y),_F2(y));`

$$\_F2(y) = \_C1 \sin(\sqrt{\_c_1}\, y) + \_C2 \cos(\sqrt{\_c_1}\, y)$$

Suppose we wish to solve the partial differential equation with the boundary conditions $u(x,0) = 0$, $u(x,b) = 0$, $u(0,y) = 0$, and $u(a,y) = f(y)$. Then we successively substitute these boundary conditions in and solve for the coefficients. We begin with the condition at $y = 0$

> `simplify(subs(y=0,_C1*sin(sqrt(_c[1])*y)+_C2*cos(sqrt(_c[1])*y)));`

$$\_C2$$

and conclude that $C_2 = 0$. Hence, the general form of the separated solution is $C_1 \sin(\sqrt{c_1}\, y)$. Next we turn to the boundary condition at $y = b$ and obtain

> `simplify(subs(y=b,_C1*sin(sqrt(_c[1])*y)+_C2*cos(sqrt(_c[1])*y)));`

$$\_C1 \sin(\sqrt{\_c_1}\, b) + \_C2 \cos(\sqrt{\_c_1}\, b)$$

In order to satisfy the boundary condition at $y = b$ the value of c_1 is chosen to be $\frac{n^2\pi^2}{b^2}$.

> `_c[1]=(n*Pi)^2/b^2;`

$$\_c_1 = \frac{n^2\pi^2}{b^2}$$

The form of $F_1(x)$ must take the form

> `_F1(x) = _C1*sinh(n*Pi/b*x);`

$$\_F1(x) = \_C1 \sinh(\frac{n\pi x}{b})$$

Hence, the general form of the solution is

> `u(x,y):=sum(c[n]*sinh(n*Pi/b*x)*sin(n*Pi/b*y),n=1..infinity);`

$$u(x,\,y) := \sum_{n=1}^{\infty} c_n \sinh(\frac{n\pi x}{b}) \sin(\frac{n\pi y}{b})$$

We wish to compute the Fourier coefficients to account for the boundary condition $u(a,y) = f(y)$. We make use of the orthogonality properties of the trigonometric functions to do this. To this end let us declare that n and k are integers.

> `assume(n, integer):assume(k, integer):`

The Fourier coefficients are then computed by

> `int(sin(k*Pi*y/b)*sin(k*Pi*y/b),y=0..b);`

$$\frac{1}{2}\, b$$

Hence we obtain

> `c[k]*1/2*b=int(f(y)*sin(k*Pi*y/b),y=0..b);`

$$\frac{1}{2}\, c_k\text{-}\, b = \int_0^b f(y) \sin(\frac{\tilde{k}\,\pi y}{b})\, dy$$

The solution then may be represented as the Fourier series

$$u := (x, y) \mapsto 2 \frac{1}{b} \sum_{n=1}^{\infty} \left(1 \int_0^b f(y) \sin\left(\frac{k\pi y}{b}\right) dy \sinh\left(\frac{n\pi x}{b}\right) \sin\left(\frac{n\pi y}{b}\right) \left(\sinh\left(\frac{n\pi a}{b}\right)\right)^{-1} \right)$$

For the next example we consider the Laplace equation written in polar coordinates, namely

```
>  pde2:=diff(u(r,theta),r$2)+1/r*diff(u(r,theta),r)
>  +1/r^2*diff(u(r,theta),theta$2);
```

$$pde2 := \left(\frac{\partial^2}{\partial r^2} u(r, \theta)\right) + \frac{\frac{\partial}{\partial r} u(r, \theta)}{r} + \frac{\frac{\partial^2}{\partial \theta^2} u(r, \theta)}{r^2}$$

pdsolve does not provide a separated solution; hence, we substitute $u(r, \theta) = R(r)F(\theta)$ into the differential and perform the separation.

```
>  expand(subs(u(r,theta)=R(r)*F(theta),pde2)/(R(r)*F(theta)));
```

$$\frac{\frac{\partial^2}{\partial r^2} R(r)}{R(r)} + \frac{\frac{\partial}{\partial r} R(r)}{R(r) r} + \frac{\frac{\partial^2}{\partial \theta^2} F(\theta)}{F(\theta) r^2}$$

The ode for $R(r)$ becomes

```
>  ode1:=expand(r^2*R(r)*(diff(R(r),r$2)/R(r)+diff(R(r),r)/
>  (R(r)*r))-c^2*R(r));
```

$$ode1 := r^2 \left(\frac{\partial^2}{\partial r^2} R(r)\right) + r \left(\frac{\partial}{\partial r} R(r)\right) - c^2 R(r)$$

We now use dsolve to solve the ordinary differential equation

```
>  dsolve(ode1,R(r));
```

$$R(r) = \_C1 \, r^c + \_C2 \, r^{-c}$$

In order to simplify this expression we relace the hyperbolic functions with exponential functions using convert

```
>  convert(_C1*sinh(c*ln(r))+_C2*cosh(c*ln(r)),exp);
```

$$\_C1 \left(\frac{1}{2} e^{(c \ln(r))} - \frac{1}{2} \frac{1}{e^{(c \ln(r))}}\right) + \_C2 \left(\frac{1}{2} e^{(c \ln(r))} + \frac{1}{2} \frac{1}{e^{(c \ln(r))}}\right)$$

Another way to pick out the dependency on powers of r would be to use the arument series in dsolve, which tells us there are two solutions r^c and r^{-c}.

```
>  dsolve(ode1,R(r),series);
```

$$R(r) = \_C1 \, r^{(-c)} \left(1 + O(r^6)\right) + \_C2 \, r^c \left(1 + O(r^6)\right)$$

The θ differential equation is

```
>  ode2:=diff(F(theta),theta$2))+c^2*(F(theta));
```

$$ode2 := \left(\frac{\partial^2}{\partial \theta^2} F(\theta)\right) + c^2 F(\theta)$$

The solutions to the θ differential equation are the sine and cosine functions.

```
>  dsolve(ode2,F(theta));
```

$$F(\theta) = \_C1 \sin(c\,\theta) + \_C2 \cos(c\,\theta)$$

Consequently, we arrive at the following Fourier series representation for the solution.

$$u := (r, \theta) \mapsto 1/2\, C_0 + \sum_{n=1}^{\infty} r^n \left(C_n \cos(n\theta) + D_n \sin(n\theta) \right)$$

The Fourier coefficients C_n and D_n may be obtained as follows

```
>  a^n*C[n]=1/Pi*int(f(theta)*cos(n*theta),theta=0..2*Pi);
```

$$a^{n\tilde{}}\, C_{n\tilde{}} = \frac{\displaystyle\int_0^{2\pi} f(\theta) \cos(n\tilde{}\,\theta)\, d\theta}{\pi}$$

```
>  a^n*D[n]=1/Pi*int(f(theta)*sin(n*theta),theta=0..2*Pi);
```

$$a^{n\tilde{}}\, D_{n\tilde{}} = \frac{\displaystyle\int_0^{2\pi} f(\theta) \sin(n\tilde{}\,\theta)\, d\theta}{\pi}$$

We are now in the position to write a procedure for computing the solution to the Laplace equation in the disk of radius a where the boundary data is of the form $u(a, \theta) = f(\theta)$.

```
>  solution_2:=proc(f,r,theta,a,N)
>  local k, j:
>  C(0):=1/a^n*int(f, theta=0..2*Pi);
>  for k from 1 to N do
>  C(k):=1/a^k*int(f*cos(k*theta),theta=0..2*Pi)/Pi;
>  D(n):=1/(a^k*Pi)*int(f*sin(k*theta),theta=0..2*Pi);
>  od;
>  u(r,theta):=C(0)/2+sum(r^j*(C(j)*cos(j*theta)+D(j)*sin(j*theta)),
>  j=1..N): end;
```

$solution\_2 := \mathbf{proc}(f, r, \theta, a, N)$
$\mathbf{local}\, k, j;$
$\quad C(0) := \mathrm{int}(f, \theta = 0..2 * \pi)/a^n\, ;$
$\quad \mathbf{for}\, k\, \mathbf{to}\, N\, \mathbf{do}$
$\quad\quad C(k) := \mathrm{int}(f * \cos(k * \theta),\, \theta = 0..2 * \pi)/(a^k * \pi)\, ;$
$\quad\quad D(n) := \mathrm{int}(f * \sin(k * \theta),\, \theta = 0..2 * \pi)/(a^k * \pi)$
$\quad \mathbf{end\ do};$
$\quad u(r, \theta) := 1/2 * C(0) + \mathrm{sum}(r^j * (C(j) * \cos(j * \theta) + D(j) * \sin(j * \theta)),\, j = 1..N)$
$\mathbf{end\ proc}$

We test this for the boundary data $f(\theta) = \theta^2$, where $a = 1$, and we take 4 terms in the series.

```
>  sol1:=solution_2(theta^2,r,theta,1,4);
```

$$sol1 := \frac{4}{3}\pi^3 + 4\, r \cos(\theta) + r^2 \cos(2\,\theta) + \frac{4}{9} r^3 \cos(3\,\theta) + \frac{1}{4} r^4 \cos(4\,\theta)$$

Let us now take more terms and plot the result

```
>  with(plots):
```

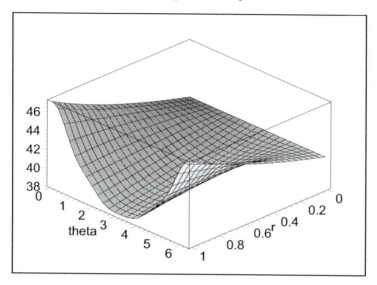

We investigate Laplace's equation in a semi-infinite, circular cylindrical region in this section. We assume the cylinder is coaxial with the z-axis and that the radius of the cylinder is b. Moreover, we assume that the cylinder exists in the region $z > 0$. First we input PDEtools.

```
>   with(PDEtools):
```

Then we input Laplace's equation in cylindrical coordinates.

```
>   pde3:=diff(v(r,theta,z),r$2)+1/r*diff(v(r,theta,z),r)
>   +1/r^2*diff(v(r,theta,z),theta$2)+diff(v(r,theta,z),z$2);
```

$$pde3 := (\tfrac{\partial^2}{\partial r^2} \mathrm{v}(r,\,\theta,\,z)) + \frac{\tfrac{\partial}{\partial r} \mathrm{v}(r,\,\theta,\,z)}{r} + \frac{\tfrac{\partial^2}{\partial \theta^2} \mathrm{v}(r,\,\theta,\,z)}{r^2} + (\tfrac{\partial^2}{\partial z^2} \mathrm{v}(r,\,\theta,\,z))$$

As pdsolve does not yield anything we try the separation of variables method, by substituting $u(r,\theta,z) := R(r)T(\theta)Z(z)$ into the equation.

```
>   expand(subs(v(r,theta,z)=R(r)*Z(z)*T(theta),pde3/
>   (R(r)*Z(z)*T(theta))));
```

$$\frac{\tfrac{\partial^2}{\partial r^2} \mathrm{R}(r)}{\mathrm{R}(r)} + \frac{\tfrac{\partial}{\partial r} \mathrm{R}(r)}{\mathrm{R}(r)\,r} + \frac{\tfrac{\partial^2}{\partial \theta^2} \mathrm{T}(\theta)}{\mathrm{T}(\theta)\,r^2} + \frac{\tfrac{\partial^2}{\partial z^2} \mathrm{Z}(z)}{\mathrm{Z}(z)}$$

It is easy to pick off the separated θ equation from the last output line. We choose the separation constant as λ^2 as we want to get sines and cosines because of periodicity of the solution.

```
>   ode1:=diff(T(theta),theta$2)+(lambda^2)*T(theta);
```

$$ode1 := (\tfrac{\partial^2}{\partial \theta^2} \mathrm{T}(\theta)) + \lambda^2 \mathrm{T}(\theta)$$

What remains after the θ equation is pulled off is

```
>   diff(R(r),r$2))/R(r)+diff(R(r),r)/(R(r)*r)+diff(Z(z),z$2))/Z(
>   z)-lambda^2/r^2;
```

$$\frac{\tfrac{\partial^2}{\partial r^2} \mathrm{R}(r)}{\mathrm{R}(r)} + \frac{\tfrac{\partial}{\partial r} \mathrm{R}(r)}{\mathrm{R}(r)\,r} + \frac{\tfrac{\partial^2}{\partial z^2} \mathrm{Z}(z)}{\mathrm{Z}(z)} - \frac{\lambda^2}{r^2}$$

We can separate this further by removing the radially dependent equation

```
>    ode2:=r^2*diff(R(r),r$2)+r*diff(R(r),r)-lambda^2*R(r)
>    +mu^2*r^2*R(r);
```

$$ode2 := r^2 \left(\tfrac{\partial^2}{\partial r^2} R(r)\right) + r \left(\tfrac{\partial}{\partial r} R(r)\right) - \lambda^2 R(r) + \mu^2 r^2 R(r)$$

Finally we have the z dependent separated solution.

```
>    ode3:=diff(Z(z),z$2)-mu^2*Z(z);
```

$$ode3 := \left(\tfrac{\partial^2}{\partial z^2} Z(z)\right) - \mu^2 Z(z)$$

We now use dsolve to solve the separated equations.

```
>    dsolve(ode1,T(theta));
```

$$T(\theta) = \_C1 \sin(\lambda \theta) + \_C2 \cos(\lambda \theta)$$

In order for $T(\theta)$ to be periodic λ must be an integer.

```
>    T_1(theta) :=subs(lambda=n,
>    _C1*cos(lambda*theta)+_C2*sin(lambda*theta));
```

$$T\_1(\theta) := \_C1 \cos(n \theta) + \_C2 \sin(n \theta)$$

We then solve for $Z(z)$

```
>    dsolve(ode3,Z(z));
```

$$Z(z) = \_C1\, e^{\mu z} + \_C2\, e^{-\mu z}$$

It is clear that we might use exponential functions instead to express the general solution $Z(z)$

$$Z(z) = \_C1 \exp(-\mu z) + \_C2 \exp(\mu z)$$

the We use the fact that $\lambda, = n$, an integer, in the radial equation, which we recognize as Bessel's equation

```
>    subs(lambda=n,dsolve(ode2,R(r)));
```

$$R(r) = \_C1\, J_n(\mu r) + \_C2\, Y_n(\mu r)$$

If the solution is to remain finite at $r = 0$ then $\backslash\_C1 = 0$, If the solution vanishes at $r = b$ then we need to compute the zeros, $\mu_1, \mu_2, \ldots$, of $J_0(\mu b)$. MAPLE has the capability of doing this as we see below.

```
>    for i from 1 to 5 do g(i):=evalf(BesselJZeros(0,i)) od;
```

$$g(1) := 2.404825558$$

$$g(2) := 5.520078110$$

$$g(3) := 8.653727913$$

$$g(4) := 11.79153444$$

$$g(5) := 14.93091771$$

If we consider only axially symmetric solutions, the situation simplifies considerably. We then have the following procedure for solving approximately the axially symmetric, boundary value problem, where the boundary condition on $z = 0$ is given by $u(r, \theta, 0) = f(r)$, the condition on the lateral sides is $u(b, \theta, z) = 0$, and the solution remains bounded as $z \to 0$.

```
>    v:= proc(f,r,b,z,N) local i, j,k, m; for
>    i from 1 to N do g(j):=evalf(BesselJZeros(0,j)) od; for k from 1
>    to N do c(k):=2*evalf(int(f*BesselJ(0,g(k)*r/b)
>    *r,r=0..b)/(b^2*(BesselJ(1,g(k))^2)))
>    od;
>    sum(BesselJ(0,g(m)*r/b)* c(m)*exp(-g(m)*z),m=1..N) end;
```

$v := \mathbf{proc}(f, r, b, z, N)$
$\mathbf{local}\ i, j, k, m;$
$\quad \mathbf{for}\ i\ \mathbf{to}\ N\ \mathbf{do}\ g(j) := \text{evalf}(\text{BesselJZeros}(0, j))\ \mathbf{od}\,;$
$\quad \mathbf{for}\ k\ \mathbf{to}\ N\ \mathbf{do}$
$\qquad c(k) := 2 \times \text{evalf}(\text{int}(f \times \text{BesselJ}(0,\ g(k) \times r/b) \times r,\ r = 0..b)/(b^2 \times \text{BesselJ}(1,\ g(k))^2))$
$\quad \mathbf{od};$
$\quad \text{sum}(\text{BesselJ}(0,\ g(m) \times r/b) \times c(m) \times \exp(-g(m) \times z),\ m = 1..N)$
$\mathbf{end}$

Let us test the procedure on the problem where $f(r) := r^2$, the radius of the cylinder is taken to be 1, and we truncate the problem with 4 terms. We obtain

```
>    v(r^2,r,1,z,4);
```

$$0.4939524358\,\text{BesselJ}(0,\ 2.404825558\,r)\,e^{(-2.404825558\,z)}$$
$$-\,0.9250217532\,\text{BesselJ}(0,\ 5.520078110\,r)\,e^{(-5.520078110\,z)}$$
$$+\,0.8059227218\,\text{BesselJ}(0,\ 8.653727913\,r)\,e^{(-8.653727913\,z)}$$
$$-\,0.7086543384\,\text{BesselJ}(0,\ 11.79153444\,r)\,e^{(-11.79153444\,z)}$$

We add another term and then plot the solution

```
>    pl1:=v(r^2,r,1,z,5):
>    with(plots):
```

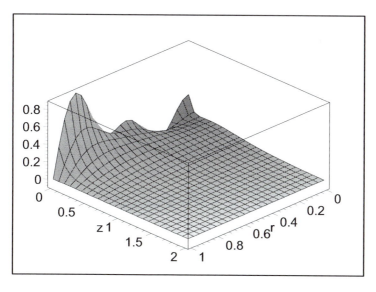

Chapter 12

Transmutations

12.1 The Method of Ascent

In this chapter we will see how we may use the idea of a transmutation to solve certain partial differential equations with variable coefficients [2,6,7]. We consider first, a particularly simple transmutation, where the differential equation's coefficients depend only on the radius, $r := \|\mathbf{x}\|$, and $\mathbf{x} \in \mathbb{R}^2$.

$$\Delta u(x,y) + a(r)r\frac{\partial u}{\partial r} + c(r)u = 0, \tag{12.1}$$

If we make the substitution

$$u(x,y) = w(x,y)\, e^{-\frac{1}{2}\int\limits_0^r a(r)r\,dr},$$

the original equation simplifies to

$$\Delta u - P(r)\,u = 0, \tag{12.2}$$

where

$$P(r) = \frac{r}{2}\partial_r a + a + \frac{r^2}{4}a^2 - c$$

The solutions of these functions may be written in terms of a **transmutation** of a solution of Laplace's equation [6,7]

$$\Delta h(\mathbf{x}) = 0 \tag{12.3}$$

in the form

$$u(\mathbf{x}) = (\mathcal{G}f)\,(\mathbf{x}) := h(\mathbf{x}) + \int\limits_0^1 \sigma G(r, 1-\sigma^2)h(\mathbf{x}\sigma^2)\,d\sigma, \tag{12.4}$$

where the integration kernel G satisfies the Goursat problem [2] [1]

$$2(1-t)\frac{\partial^2 G}{\partial t \partial r} - \frac{\partial G}{\partial r} + r\left(\frac{\partial^2 G}{\partial^2 r} - P(r^2)G\right) = 0$$

$$G(r,0) = 0, \quad G(r,0) + \int_0^r P(t^2)t\,dt. \tag{12.5}$$

The operator, $\mathcal{G}$, has been referred to by Robert Carrol [2] as the Bergman-Gilbert operator, perhaps more correctly would be the Bergman-Gilbert-Vekua operator. The question the reader might ask now is "For what reason do we want to solve the equation for the G-function, which is more difficult that the equation 12.1?" The answer is that once we have the G-function we may find *any* solution from the transmutation of a complete family of functions, for example the harmonic polynomials form such a family. The above representation result can be easily extend to $\mathbb{R}^n$, i.e. to equations of the form

$$\frac{\partial^2 u}{\partial x_1^2} + \cdots + \frac{\partial u}{\partial x_n^2} - P(r)u = 0; \tag{12.6}$$

in this case the representation becomes (12.7)

$$u(\mathbf{x}) = (\mathcal{G}f)(\mathbf{x}) := h(\mathbf{x}) + \int_0^1 \sigma^{n-1}G(r,1-\sigma^2)h(\mathbf{x}\sigma^2)\,d\sigma, \tag{12.7}$$

where we notice in the case of $\mathcal{R}^2$ we replace 2 by n in (12.7) to obtain (12.4) [6, 7]. We notice that nowhere in the equation for the G-kernel does the dimension n appear; hence, the name "method of ascent". The harmonic functions, $\mathfrak{H}$, form a space of two-time, continuous solutions of (12.3). If we call the two-time, continuous solutions of (12.2) $\mathfrak{M}$ then $\mathcal{G}$ represents a mapping of the space $\mathfrak{H}$ into the space $\mathfrak{M}$, i.e

$$\mathcal{G} : \mathfrak{H} \to \mathfrak{M}. \tag{12.8}$$

The question now is whether $\mathcal{G}$ has an inverse, $\mathcal{G}^{-1}$. If this is the case, then the mapping $\mathcal{G}$ is one-to-one **onto** $\mathfrak{H}$, and we know that each solution of (12.2) corresponds to one-and-only-one harmonic function. It is easy to find a series solution for the kernel $G(r,t)$; namely, we seek a solution in the form

$$G(r,t) = \sum_{n=1}^{\infty} g_n(r^2)t^{n-1}. \tag{12.9}$$

---

[1] A Goursat boundary condition is given on the characteristics of the differential equation.

Upon substituting the series (12.9) into the differential equation for the function $G(r,t)$,(12.5), we obtain in the usual manner the recursion formulae

$$g_1(r) = -\int_0^r rP(r)\,dr,$$

and for $k = 1, 2 \cdots$

$$g_{k+1}(r) = -\frac{1}{2k}\int_0^r r\frac{d^2 P(r)}{dr^2}\,dr + (2k-1)\frac{dg_k(r)}{dr} + rP(r)g_k(r).$$

However, we may use this method even when $P(r)$ has two continuous derivatives. In several simple cases the kernel may be explicitly calculated in closed form. We offer a few examples below. If $P(r) := k^2$, then

$$G(r,t) = kr\frac{J_1(krt^{\frac{1}{2}})}{krt^{\frac{1}{2}}},$$

where $J_1(x)$ is a Bessel function of the first kind and order one. If $P(r) = k^2 r^{2(m-1)}$, $M = 2, 3, \cdots$, then

$$G(r,t) = -kr^m t^{m/2-1} J_1\left(\frac{kr^m}{m}t^{m/2}\right)$$

If $P(r) := \frac{4k(k+1)}{(1+r^2)^2}$, then

$$G(r,t) = -\frac{2k(k+1)r^2}{1+r^2}F\left(k+2, 1-k; 2; \frac{r^2 t}{1+r^2}\right),$$

where $F(\alpha, \beta; \gamma; z)$ is the hypergeometric series. Let us see how MAPLE can be used to construct a G-function.

```
GfuncTaylor := proc (F, N)
local k;
description "This is a program to generate the Taylor coefficients of the G
function";
    g(1) := int(F * r, r);
    for k to N do
        g(k+1) := -1/2*k^-1*int(r*diff(c(l), r, r)+(2*k+1)*diff(g(k), r)+
F * g(k), r)
    end do;
end proc;
```

GfuncTaylor := **proc** (F, N)
local k;
description "This is a program to generate the Taylor coefficients of the G function";
    $g(1) := int(F * r, r)$;
    for k **to** N **do**
        $g(k+1) := -1/2 * k\hat{\ }-1 * int(r * diff(g(l), r, r) + (2*k+1) * diff(g(k), r)$

$+ F * g(k), r)$
    end do;
end proc;
GfuncTaylor $(r^2, 6)$

$$\frac{3003\,r^4}{4096} + \frac{2689\,r^7}{26880} + \frac{48259\,r^{10}}{12902400} + \frac{19\,r^{13}}{349440} + \frac{37\,r^{16}}{107347968} + \frac{r^{19}}{1062297600} + \frac{r^{22}}{1121786265600}$$

This program lists the last Taylor coefficient of the *G* function.

Gfunction := **proc** (FN)
local k, l;
description "This is a program to generate the G function to N terms.";
    GfuncTaylor(F, N);
    $sum(c(l) * t\hat{\ }l, l = 1..N)$
end proc;

Gfunction := **proc** (FN)

local k, l;
description "This is a program to generate the G function to N terms.";
    GfuncTaylor(F, N);
    $sum(c(l) * t\hat{\ }l, l = 1..N)$
end proc;

Gfunction $(r^2, 4)$

$$\frac{r^4 t}{4} + \left(-\frac{3\,r^4}{8} - \frac{r^7}{56}\right) t^2 + \left(\frac{15\,r^4}{32} + \frac{r^7}{28} + \frac{r^{10}}{2240}\right) t^3 + \left(-\frac{35\,r^4}{64} - \frac{71\,r^7}{1344} - \frac{r^{10}}{896} - \frac{r^{13}}{174720}\right) t^4$$

$$plot3d\left(\frac{r^2t}{2}+\left(-\frac{3\,r^2}{4}-\frac{r^3}{12}\right)t^2+\left(\frac{15\,r^2}{16}+\frac{r^3}{6}+\frac{r^4}{192}\right)t^3\right.$$
$$\left.+\left(-\frac{35\,r^2}{32}-\frac{71\,r^3}{288}-\frac{5\,r^4}{384}-\frac{r^5}{5760}\right)t^4,r=0\ldots2,t=0\ldots3\right)$$

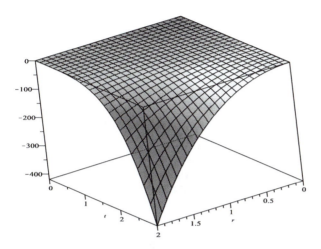

TransmutationG := **proc** (FNH)
local u, g;
description "This is a program to generate the G transmutation of a harmonic function to u,
  having N terms in the G-gunction";
    g := subs(t = −s^2 + 1, *Gfunction*(F, N));
    u := H + int(s * g * subs(r = r * s^2, H), s = 0..1)
end proc;

TransmutationG := **proc** (FNH)
local u, g;
description "This is a program to generate the G transmutation
of a harmonic function to u, having N terms in the G-gunction";
    g := subs(t = −s^2 + 1, *Gfunction*(F, N));
    u := H + int(s * g * subs(r = r * s^2, H), s = 0..1)
end proc;
TransmutationG $(1, 4, r\cos(\theta))$
$evalc\left(\mathrm{Re}\left((x+iy)^3\right)\right)$

$$x^3-3\,xy^2$$

$evalc\left(\mathrm{Im}\left((x+iy)^3\right)\right)$

$$3\,x^2 y - y^3$$

$$h\,(2) := x$$
$$h\,(4) := x^2 - y^2$$
$$h\,(6) := x^3 - 3\,xy^2$$
$$h\,(8) := x^4 - 6\,x^2 y^2 + y^4$$
$$h\,(10) := x^5 - 10\,x^3 y^2 + 5\,xy^4$$
$$h\,(12) := x^6 - 15\,x^4 y^2 + 15\,x^2 y^4 - y^6$$
$$h\,(14) := x^7 - 21\,x^5 y^2 + 35\,x^3 y^4 - 7\,xy^6$$
$$h\,(16) := x^8 - 28\,x^6 y^2 + 70\,x^4 y^4 - 28\,x^2 y^6 + y^8$$
$$h\,(18) := x^9 - 36\,x^7 y^2 + 126\,x^5 y^4 - 84\,x^3 y^6 + 9\,xy^8$$
$$h\,(20) := x^{10} - 45\,x^8 y^2 + 210\,x^6 y^4 - 210\,x^4 y^6 + 45\,x^2 y^8 - y^{10}$$

$$h\,(1) := y$$
$$h\,(3) := 2\,xy$$
$$h\,(5) := 3\,x^2 y - y^3$$
$$h\,(7) := 4\,x^3 y - 4\,xy^3$$
$$h\,(9) := 5\,x^4 y - 10\,x^2 y^3 + y^5$$
$$h\,(11) := 6\,x^5 y - 20\,x^3 y^3 + 6\,xy^5$$
$$h\,(13) := 7\,x^6 y - 35\,x^4 y^3 + 21\,x^2 y^5 - y^7$$
$$h\,(15) := 8\,x^7 y - 56\,x^5 y^3 + 56\,x^3 y^5 - 8\,xy^7$$
$$h\,(17) := 9\,x^8 y - 84\,x^6 y^3 + 126\,x^4 y^5 - 36\,x^2 y^7 + y^9$$
$$h\,(19) := 10\,x^9 y - 120\,x^7 y^3 + 252\,x^5 y^5 - 120\,x^3 y^7 + 10\,xy^9$$

$$h\,(2) := r\cos\,(\theta)$$
$$h\,(4) := -\,(\sin\,(\theta))^2\,r^2 + (\cos\,(\theta))^2\,r^2$$
$$h\,(6) := -3\,(\sin\,(\theta))^2\,\cos\,(\theta)\,r^3 + (\cos\,(\theta))^3\,r^3$$
$$h\,(8) := (\sin\,(\theta))^4\,r^4 - 6\,(\sin\,(\theta))^2\,(\cos\,(\theta))^2\,r^4 + (\cos\,(\theta))^4\,r^4$$
$$h\,(10) := 5\,(\sin\,(\theta))^4\,\cos\,(\theta)\,r^5 - 10\,(\sin\,(\theta))^2\,(\cos\,(\theta))^3\,r^5 + (\cos\,(\theta))^5\,r^5$$
$$h\,(12) := -\,(\sin\,(\theta))^6\,r^6 + 15\,(\sin\,(\theta))^4\,(\cos\,(\theta))^2\,r^6 - 15\,(\sin\,(\theta))^2\,(\cos\,(\theta))^4\,r^6 + (\cos\,(\theta))^6\,r^6$$

We have omitted some of the output below, as it is too large to fit the page.

$$h\,(1) := r\sin\,(\theta)$$

$$h\,(3) \; := \; 2\,\sin\,(\theta)\,\cos\,(\theta)\,r^2$$

$$h\,(5) \; := \; -\,(\sin\,(\theta))^3\,r^3 + 3\,\sin\,(\theta)\,(\cos\,(\theta))^2\,r^3$$

$$h\,(7) \; := \; -4\,(\sin\,(\theta))^3\,\cos\,(\theta)\,r^4 + 4\,\sin\,(\theta)\,(\cos\,(\theta))^3\,r^4$$

$$h\,(9) \; := \; (\sin\,(\theta))^5\,r^5 - 10\,(\sin\,(\theta))^3\,(\cos\,(\theta))^2\,r^5 + 5\,\sin\,(\theta)\,(\cos\,(\theta))^4\,r^5$$

$$h\,(11) \; := \; 6\,(\sin\,(\theta))^5\,\cos\,(\theta)\,r^6 - 20\,(\sin\,(\theta))^3\,(\cos\,(\theta))^3\,r^6$$
$$+\,6\,\sin\,(\theta)\,(\cos\,(\theta))^5\,r^6$$

$$h\,(13) \; := \; -\,(\sin\,(\theta))^7\,r^7 + 21\,(\sin\,(\theta))^5\,(\cos\,(\theta))^2\,r^7$$
$$-\,35\,(\sin\,(\theta))^3\,(\cos\,(\theta))^4\,r^7 + 7\,\sin\,(\theta)\,(\cos\,(\theta))^6\,r^7$$

The harmonic functions may then be transmuted into the family of solutions of the original equation. A particular solution may be approximated using these functions. We may use as many solutions as we need; however, the maple results should be repressed not to use up too much screen space. The method may be extended to higher-order differential equations involving products of the Laplacian operator such as

$$\Delta^2 u(\mathbf{x}) + A(r^2)\Delta u(\mathbf{x}) + B(r^2)u(\mathbf{x}) = 0, \quad \mathbf{x} \in \mathbb{R}^2 \tag{12.10}$$

In [3] the method considered to $\mathbf{x} \in \mathbb{R}^n$, $n = 2, 3, \cdots$; however, here we restrict our examples to the case of $\mathbb{R}^2$. In the paper [3] the fourth-order differential equation

$$u(\mathbf{x}) = h^{(1)}(\mathbf{x}) + r^2 h^{(1)}(\mathbf{x}) + \int_0^1 \sigma^{n-1} G^{(1)}(r^2, 1-\sigma^2) h^{(1)} \tag{12.11}$$

was considered. It was found that the ideas used for refgilbert-operator could be extended and solutions were found in the form $u(\mathbf{x}) := h^{(1)}(\mathbf{x}) + h^{(2)}(\mathbf{x}) +$
$\int_0^1 \sigma G^{(1)}(r^2, 1-\sigma^2) h^{(1)}(\mathbf{x}\sigma^2)\, d\sigma +$

$$\int_0^1 \sigma G^{(2)}(r^2, 1-\sigma^2) h^{(2)}(\mathbf{x}\sigma^2)\, d\sigma \tag{12.12}$$

The kernel functions $G^{(i)}(r^2, t)\, i = 1, 2$ satisfy the partial differential equation

$$G_{rrrr} + \frac{4(1-t)}{r} G_{rrrt} + \frac{4(1-t^2)}{r^2} G_{rrt} - \frac{12(1-t)}{r^2} G_{rrt}$$

$$-\frac{4(1-t)^2}{r^3} G_{rtt} + \frac{2(1-t)}{r}\left[\frac{6}{r^2} + A(r^2)\right] G_{rt} - \frac{2}{r} G_{rrr}$$

$$+\left[\frac{3}{r^2} + A(r^2)\right] G_{rr} - \frac{1}{r}\left[\frac{3}{r^2} + A(r^2)\right] G_r + B(r^2) G = 0 \tag{12.13}$$

The kernel function $G^{(1)}(r, t)$ satisfies the Goursat conditions

$$G^{(1)}(0, t) = 0, \quad \lim_{r \to 0} G_r^{(1)}(r, t) = 0 \tag{12.14}$$

$$G^{(1)}(r,0) = 0, \quad G^{(1)}_{rrt}(r,0) - \frac{1}{r}G^{(1)}_{rt}(r,0) + \frac{r^2}{2}B(r^2) = 0; \qquad (12.15)$$

whereas $G^{(2)}(r,t)$ satisfies

$$G^{(2)}(0,t) = 0, \quad \lim_{r\to 0} G^{(2)}_r(r,t) = -2,$$

$$G^{(2)}_{rr}(r,0) - \frac{1}{r}G^{(2)}_{rt} + 2r^2 A(r^2) = 0,$$

$$G^{(2)}_{rrt}(r,0) - \frac{1}{r}G^{(2)}_{rt}(r,0) + rG^{(2)}_{rrr}(r,0) - \left[G^{(2)}_{rr}(r,0) - \frac{1}{r}G^{(2)}(r,0) \right]$$

$$+ 2r^2 A(r^2) + \frac{r}{2}A(r^2)G^{(2)}_r(r,0) + \frac{r^4}{2}B(r^2) = 0 \qquad (12.16)$$

12.2   Orthogonal Systems of Functions

The integral

$$(g,h) := \int_{\mathcal{D}} g(\mathbf{x})h(\mathbf{x})\,dx \text{ where } \mathcal{D} \subset \mathbb{R}^2 \qquad (12.17)$$

and where $\mathcal{D}$ is a finite domain, is called the inner product of the functions $g(\mathbf{x})$ and $h(\mathbf{x})$. If h and g are the same then the integral

$$\|h\|^2 := (h,h) = \int_{\mathcal{D}} h(\mathbf{x})^2\,dx \qquad (12.18)$$

is called the norm of h. Moreover, if the norm is equal to 1, we say the function h is normalized. If a family of normalized functions $f_i(\mathbf{x})$, $i = 1, 2, 3 \cdots$, such that any two functions are orthogonal is called an orthonormal family, ie it satisfies the relations

$$(f_i, g_j) = \delta_{ij}, \quad \text{where} \quad \delta_{ii} = 1, \text{but } \delta_{i,j} = 0 \text{ for} i \neq j. \qquad (12.19)$$

These definitions may be extended to complex functions where we now say two functions are orthogonal when infinite family of functions $(f, \bar{g}) = (g, \bar{f}) = 0$, where a bar over a function indicates the complex conjugate. The norm of a function is then given by $\|f\| = (f, \bar{f})$ .istx Definition: We say a finite collection of r functions is linear dependent if there exist constants c_i not all zero such that $c_1 f_1(\mathbf{x}) + c_2 f_2(\mathbf{x}) + \cdots + c_r f_r(\mathbf{x}) = 0$ for all values of x. Otherwise, the collection is dependent. If we have an infinite family of functions, $\mathcal{F}$, such that for any n of functions is linearly independent an orthonormal set may may be obtained. This may be accomplished as follows: The first function is chosen as $\varphi_1 := \frac{f_1}{\|f_1\|}$. To compute the second member of the orthonormal set, we seek a number α_1, such that the function $f_2 - \alpha_1$ is orthogonal φ_1. We are led to the equation $(f_2 - \alpha_1, \varphi_1) = (f_2, \varphi_1) - \alpha_1 = 0$, or $\alpha_1 = (f_2, \varphi_1)$, or $\varphi_2 = f_2 - (f_2, \varphi_1)\varphi_1$. To find φ_3 we seek two constants α_1' and α_2', such that $f_3 - \alpha_1'\varphi_1 - \alpha_2'\varphi_2$ is perpendicular to both φ_1 and φ_2. This leads to $\varphi_3 = f_3 - (f_3, \varphi_1)\varphi_2 -$ It is clear how to continue this process.

The coefficients $c_k := (f, \varphi_k)$ are known as the expansion coefficients of the function f with respect to the orthonormal family $\{\varphi_k\}$. We should like to approximate in some way $f(x) \approx \sum_{k=0}^{\infty} c_k \varphi_k(x)$ There are examples of why this is not posisible in a point wise sense [4]; however, it is possible to form an approximation in the mean, i.e. we can show under fairly broad conditions [4] that the mean-square error $\int_{\mathcal{D}} \left(f - \sum_{k=1}^{n} c_k \varphi \right)^2 d\mathbf{x} \to 0$, $n \to \infty$ It is possible to orthonormalize the transmutations of the harmonic functions. Moreover, these orthonormal functions may be used to approximate solutions to the associated differential equation.

There many ways by which we may orthonormalize a family of functions, one of which is by using a boundary norm such generated by the inner-product

$$(g, h) := \int_{\mathcal{C}} g(\mathbf{x})h(\mathbf{x}) \, ds(\mathbf{x}),$$

where $\mathcal{C} = \partial\mathcal{D}$ is the boundary and $ds(\mathbf{x})$ is the arc-length differential. The boundary-norm is then given by $\|f\|^2 := (f, f)$, which would be useful in approximating Dirichlet data for a domain with $\mathcal{C}$ as its boundary.

12.3 Acoustic Propagation

Another transmutation which occurs in the theory of acoustic propagation in a wave guide. In particular, if the wave guide has an index of refraction that depends on the vertical distance from the upper surface we are interested in solving the axially symmetric Helmholtz equation.

$$\frac{\partial^2 u}{\partial r^2} + \frac{1}{r}\frac{\partial u}{\partial r} + \frac{\partial^2 u}{\partial z^2} + k^2 n^2(z)u = 0, \quad 0 \le z_b. \tag{12.20}$$

One way to generate solutions to this problem is to find a transmutation from a simpler equation; for further detains on this method we refer the reader to [9]. To illustrate the method we consider the constant coefficient Helmholtz equation

$$\frac{\partial^2 h}{\partial r^2} + \frac{1}{r}\frac{\partial h}{\partial r} + \frac{\partial^2 h}{\partial z^2} + k_0^2 h = 0, \quad 0 \le z_b. \tag{12.21}$$

Indeed, we seek a transmutation in the form

$$u(r, z) = (\mathbb{I} + \mathbb{K}) \, h(r, z) := h(r, z) + \int_{z_b}^{z} K(z, s) \, h(r, s) \, ds \qquad (12.22)$$

Recalling that $h(r, z)$ must satisfy (12.20) and substituting (12.21) it follows that $K(z, s)$ must satisfy the equation

$$\frac{\partial^2 K}{\partial z^2} - \frac{\partial^2 K}{\partial s^2} + k_0^2 \left[n^2(z) - 1 \right] K = 0, \quad (z, s) \in [0, z_b] \times (z, s) \in [0, z_b].$$
$$(12.23)$$

To prove this it is sufficient substitute the representation into (12.21) and perform two integrations by parts to obtain $0 = k^2[n^2(z) - 1]h(r, z) + \frac{\partial}{\partial z}[K(z, z)h(r, z)] + \int_{z_b}^{z} \left(K_{11} - K_{22} + k^2[n^2(zv) - 1]K \right) h(r, s) \, ds - K(z, zh_z(r, z)h_z(r, z)$

$$+ K(z, z_b)h_z(r, z_b) + K_2(z, z)h(r, z) - K_2(z, z_b)h(r, z_b) \qquad (12.24)$$

As the integral must automatically vanish for every possible harmonic function, we obtain the differential equation (12.23). To make the remaining terms vanish we must choose the appropriate Goursat conditions. This is accomplished by choosing on the characteristic $x = z$

$$2 \frac{\partial}{\partial z} K(z, z) + k_0^2[n^2(z) - 1] = 0 \qquad (12.25)$$

We have now just the term $K(z, z_b)h_z(r, z) - K_2(z, z_b)h(r, z_b)$ to be removed. From the representation (12.22) we notice that $u(r, z_b) = h(r, z_b)$ and $\frac{\partial u(r, z_b)}{\partial x} = \frac{\partial h(r, z_b)}{\partial x} + K(z_b, z_b)h(r, z_b)$; hence, $u(r, z)$ satisfies the same boundary condition as $h(r, z)$. It follows that if $u(r, z)$ satisfies the semi-soft, homogeneous, boundary condition $au(r, z_b) + b \frac{\partial u(r, z_b)}{\partial z} = 0$, then $K(z, s)$ satisfies

$$aK(z, z_b) + bK_2(z, z_b) = 0. \qquad (12.26)$$

Note that if b =0 this becomes the homogeneous Dirichlet (soft) condition; whereas, if a=o this is the homogeneous Neumann (hard) condition.

EXERCISES

Exercise 12.1. *Prove that $G(r, t)$ solves the differential equation*

$$2(1 - t) \frac{\partial^2 G}{\partial t \partial r} - \frac{\partial G}{\partial r} + r \left(\frac{\partial^2 G}{\partial^2 r} - P(r^2)G \right) = 0$$

Hint: Substitute the representation for $u(x, t)$ into the differentel equation (12.21) and integrate by parts.

Exercise 12.2. *Check the special cases of simple $P(r)$ terms, given in the text, to see if the first few terms of the corresponding $G(r, t)$ match with those you calculate using an approximate transmutation.*

Exercise 12.3. *Compute the transmuted first few harmonic functions for the special examples presented in the first section.*

Exercise 12.4. *Show that the family of trigonometric functions $\frac{1}{\sqrt{2\pi}}$, $\frac{\cos(x)}{\sqrt{\pi}}$, $\frac{\sin(x)}{\sqrt{\pi}}$,*
fraccos$(2x)\sqrt{\pi}$,
fracsin$(x)\sqrt{\pi}$, $\frac{\cos(3x)}{\sqrt{\pi}}$, $\cdots$ over the interval $[0, 2\pi]$, is orthonormal.

Exercise 12.5. *Show that the family of exponential functions*

$$\frac{1}{\sqrt{2\pi}}, \frac{e^{ix}}{\sqrt{2\pi}}, \frac{e^{2ix}}{\sqrt{2\pi}}, \frac{e^{3ix}}{\sqrt{2\pi}}, \cdots$$

over the interval $[0, 2\pi]$ is orthonormal.

Exercise 12.6. *Show that the kernel functions $G^{(i)}(r^2, t)$ $i = 1, 2$ for the equation $\Delta^2 u(\mathbf{x}) + A(r^2)\Delta u(\mathbf{x}) + B(r^2)u(\mathbf{x}) = 0$, $\mathbf{x} \in \mathbb{R}^2$ satisfy the partial differential equation (12.13)..*

Exercise 12.7. *Show that the wave propagation kernel $K(z, s)$ satisfies the partial differential equation*

$$\frac{\partial^2 K}{\partial z^2} - \frac{\partial^2 K}{\partial s^2} + k_0^2 \left[n^2(z) - 1 \right] K = 0.$$

Bibliography

[1] H. Begehr and R. P. Gilbert: *Transformations, Transmutations, and Kernel Functions vol1*, Pitman Monographs and Surveys in Pure and Applied Mathematics **58**, Longman Scientific & Technical (1992).

[2] R. Carroll; *Transmutation Theory and Applications*, North Holland, (2012).

[3] D. L. Colton and R.P. Gilbert: *Integral operators and complete families of solutions for* $\Delta_{p+2}^2 u(\mathbf{x}) + A(r^2)\Delta_{p+2} + B(r^2)u(\mathbf{x}) = 0$, Arch. Rat. Mech. Anal. **43** (1971), 62-78.

[4] R. Courant and D. Hilbert: *Methods of Mathematical Physics* vol. 1, Interscience, New York (1953).

[5] A. Erdélyi, W. Magnus, F. Oberhettinger, F. G. Tricomi: *Higher Transcendental Functions*, McGraw Hill, (1953).

[6] R. P. Gilbert: *A method of ascent for solving boundary value problems*, Bull. Amer. Math. Soc. **75** (1969).

[7] R. P. Gilbert: *The construction of solutions for boundary value problems by function theoretic methods*, SIAM Journal Math. Anal. **1** (1),96-114, (1970)

[8] R. P. Gilbert and G. C. Hsiao: MAPLE *Projects for Differential Equations*, Prentice Hall, (2002).

[9] R. P. Gilbert and D. H. Wood: *A transmutation approach to underwaterr sound propagation*, Wave Motion **8**, 383-397, (1986).

[10] R. P. Gilbert, D. H. Wood and Yonzhi Xu: *Construction of approximations to acoustic Greens functions for nonhomogeneous oceans using transmutation*, Wave Motion **10**, 285-297, (1988).

[11] W. Magnus and W. Obberhettinger: *Functions of Mathematical Physics*, Chelsea, New York, (1949).

[12] P. M. Morse and H. Feshbach: *Methods of Theoretical Physics* Part 1, McGraw Hill (1953).

Index

assign, 107
asymptotic expansion, 107
autonomous systems, 147
axially symmetric solution, 216

beats, 88
Bernoulli equation, 15
Bessel, 171
boundary value problems, 99
boundary-layer, 106
BOXED, 23, 200
BVP, 103

Catalytic reactor, 106
Cauchy condition, 192
Cauchy problem, 191, 195
central difference formulas, 101
characteristic equation, 58, 118
charmat, 115
charstrip, 195
coeff, 126
coeffs p q, 136
collect, 86
combine, 84
complex eigenvalues, 152
condition, 144
const coeff, 56
convert, 134
convolution, 187
convolution integral, 187
Cramer's rule, 77
CrPt, 149
cylindrical coordinates, 215

damped, 92
dampened pendulum, 156

damping, 154
DEtools, 1
dfieldplot, 3, 4
diffusion, 106
Digits, 97
Dirac, 171
Dirac delta function, 170
direction field, 3
discontinuous functions, 13
discontinuous right-hand side, 177
display, 4
distinct real eigenvalue, 151
dsolve, 19, 61

eigenvalue problem, 111
eigenvalues, 111, 112
equilibrium points, 148
erf, 176
error function, 177
EULER, 49
Euler method, 42
evalc, 71
evalf, 23
Exact equ, 28
exact equ, 28
exact equations, 27
exact1, 27
exponential integral, 176

finite difference method, 101
first-order linear equations, 11
first-order PDE, 194
firsteuler, 44
Fourier, 180
Fourier coefficients, 212
Fourier series representation, 213

Fourier transform, 179
fouriercos, 180
fouriersin, 179
frobenius 1, 137
frobenius 2, 140
frobenius 3, 142
frobenius 4, 144
fsolve, 123

gausselim, 105
gaussjord, 105
genmatrix, 105

Heat equation, 193, 199
heat equation 1, 200
Heaviside, 170
higher-order equations, 70
homogeneous boundary value
        problem, 111

impeuler, 44
improved Euler, 43
improved euler, 45
independent solutions, 59
indicial equation, 133
initial curve, 192
initial value problem, 3
inttrans, 173
inttransforms, 179
invfourier, 181
invlaplace, 173
IVP, 60
IVP soln, 60

Jacobi matrix, 149

Kummer functions, 176

laplace, 168, 173
Laplace equation, 193, 211
Laplace transform, 167
linalg, 101
linear systems, 149
local, 22
logistic equation, 50

map, 86
Maple
        sessions, i
mix 1, 19
mix 2, 21
mixing problems, 18
my wronskian, 57

natural frequency, 92
NEWTON, 118
Newton-Raqphson, 118
nonlinear differential equations,
        129
nonlinear pendulum, 154
normal, 107, 185
nullspace, 115
numeric, 84, 127
numerical solutions, 9
numpoints, 127

odeadvisor, 65
orbits, 148
order of convergence, 51
our picard, 34

particular solution, 75
pdsolve, 189, 211
PDtools, 191
phase plane, 148
phaseportrait, 150
Picard Iteration, 34
plain euler, 44
plots, 2
polar, 71
polar coordinates, 213
pos, 139
program:BVP, 103
program:convolution, 187
program:frobenius 2, 140
program:frobenius 3, 142
program:frobenius 4, 144
program:heat equation, 200
program:heat equation$_2$, 203
program:initial value problem, 60
program:my wronskian, 57

program:NEWTON, 118
program:snap shot string, 208
program:soln series, 161
program:soln series 2, 164
program:Vari Parama2, 62
program:wave equation$_1$, 206
programs:const coef, 56
pulse, 183
purely imaginary eigenvalues, 152

recursion formula, 136
regular singular points, 132
resonance, 90
Riccati equation, 16
runge katta, 47
Runge-Kutta, 43

save, 119
second-order equations, constant
        coefficients, 55
semi-batch reactor, 39
separable equations, 24
separable initial, 26
separablesol, 25
separation of variables, 208
series, 107
series methods, 125
simple harmonic motion, 83
singular perturbation, 110
snap shot, 208
solution 2, 214

solve, 19
solving PDE with Fourier
        transform, 185
Sturm-Liouville Problem, 111
style = POINT, 39
symbol = CROSS, 39
symbol=CIRCLE, 54

Table of Laplace transforms, 172
taylor, 125
Taylor coefficients, 126
taylor picard, 36
tooth, 183
trigsubs, 87
two-point boundary value
        problem, 103
type, 139

unapply, 107
unassume, 65
undetermined coefficients, 63, 85
undeternined coefficients, 72

Vari Param2, 62
variation of parameters, 75
vibrating string, 205

wave equation, 191
wave equation 1, 206
Wron, 79
wronsk, 79
Wronskian, 57, 77